Flyer's Recreation Guide - NW

Reed I. White

Alta Research, Lava Hot Springs Airpark, Idaho

Flyer's Recreation Guide - Northwest

Written by	Reed I. White
Published by	Alta Research
	131 NW 4th Street #290
	Corvallis, OR 97330-4702, USA
	alta@alta-research.com
	http://www.alta-research.com
Copyright	© 2001 by Alta Research. All rights reserved.

No part of this book may be reproduced or transmitted in any form or by any means, electronic or mechanical, including photocopying, recording, or by any information storage and retrieval system without written permission from the author, except for inclusion of brief quotations in a review.

Printing History First northwest edition published in 1997. Printed 2001, Second Edition, USA.

Library of Congress Number 97-93365

Cataloging Data

White, Reed I.
   Flyer's Recreation Guide - Northwest
   Access to Adventure in the Northwestern States.
   Includes index.
1. Air Pilot Guides — The Northwest.
2. Travel — Northwest (U.S.).
3. Recreation — Northwest (U.S.).
4. Northwest (U.S.). I. Title.
TL 726.W45   629.13   97-93365
ISBN 0-945549-08-3 Softcover

About This Book

The **Flyer's Recreation Guides** are a resource of ideas and information for pilots and passengers who fly for fun and adventure. This Guide contains descriptions of northwest America's most exciting sites within convenient access of landing strips. Update bulletins at *http://www.alta-research.com* keep the Guide growing and current.

All sites in the Guide have been visited and researched by experienced, fun-loving pilots. Each chapter provides detailed information on a specific site to help you create your own memorable adventures. Included are tips for hikes, swimming, fishing, lodging, restaurants, and entertainment. The Guide tells how to find trailheads, hot springs, raft trips, and other items of interest. Maps and photographs help you visualize what each site has to offer.

Airport information is provided to help determine if your aircraft and skills are compatible with the destination. Aerial photographs help you locate and identify airstrips. Local services and transportation are listed. Hundreds of phone numbers are included so that you can get further information, make reservations, and verify current conditions. Also included is information that will increase the safety and enjoyment of your trip.

We encourage you to fly safely as you use this book to create your own adventures. I would love to hear about your experiences – **feedback and suggestions are always welcome!**

R/White

Reed I. White

alta@alta-research.com
www.alta-research.com

Other sources of fly-in recreation information from Alta Research

Flyer's Recreation Guide - SW, ISBN 0-945549-09-1

The southwest version covers the states of Arizona, California, Colorado, New Mexico, Nevada, and Utah.

Web site: **http://www.alta-research.com**

The Alta Research web site contains update information, links to other aviation sites, and ways to order publications.

Recommended publications from other authors

Flight Guide by Airguide Publications
Fly Idaho! by Galen Hansen
Fly the Big Sky! by Galen Hansen
Pilot Getaways Magazine by Airventure

© 2001 by **ALTA** Research

Table of Contents

Chapter	Location	Page
Idaho		
Big Creek	Central Idaho Wilderness	**ID - BIGC**
Bruce Meadows	Central Idaho Wilderness	**ID - BRUC**
Chamberlain	Central Idaho Wilderness	**ID - CHAM**
Dixie Town	Northern Idaho Wilderness	**ID - DIXT**
Dixie USFS	Northern Idaho Wilderness	**ID - DIXU**
Flying B Ranch	Central Idaho Wilderness	**ID - FLYB**
Johnson Creek	Central Idaho Wilderness	**ID - JOHN**
Lava Hot Springs	South-eastern Idaho	**ID - LAVA**
Moose Creek	Central Idaho Wilderness	**ID - MOOS**
Murphy Hot Springs	Southern Idaho	**ID - MURP**
Silverwood Amusement Park	Northern Idaho	**ID - SILV**
Smiley Creek	Central Idaho	**ID - SMIL**
Stanley	Central Idaho	**ID - STAN**
Sulfur Springs Ranch	Idaho Wilderness	**ID - SULF**
Thomas Creek	Central Idaho Wilderness	**ID - THOM**
Warm Springs	Central Idaho Wilderness	**ID - WARM**

© 2001 by ALIA Research

Intro-6

Montana

Benchmark	NW Montana Wilderness	**MT - BENC**
Big Fork	Northwestern Montana	**MT - BIGF**
Chico Hot Springs	Southwestern Montana	**MT - CHIC**
Hot Springs	Northwestern Montana	**MT - HOTS**
Schafer	NW Montana Wilderness	**MT - SCHA**
Spotted Bear	NW Montana Wilderness	**MT - SPOT**

Oregon

Alvord Desert	Southeastern Oregon	**OR - ALVO**
Ashland	Southern Oregon	**OR - ASHL**
Bandon	Oregon Coast	**OR - BAND**
Diamond Peak	Oregon Cascade Mountains	**OR - DIAM**
Flying M Ranch	Northwestern Oregon	**OR - FLYM**
Hood River	Northern Oregon	**OR - HOOD**
Horse Ranch	NE Oregon Wilderness	**OR - HORS**
McMinnville	Northwestern Oregon	**OR - MCMI**
Memaloose	NE Oregon Wilderness	**OR - MEMA**
Minam Lodge	NE Oregon Wilderness	**OR - MINA**
Newport	Oregon Coast	**OR - NEWP**
Owyhee Reservoir	Eastern Oregon	**OR - OWYH**
Pinehurst	Southern Oregon	**OR - PINE**
Santiam Junction	Cascade Mountains	**OR - SANT**
Seaside	Oregon Coast	**OR - SEAS**

Sisters	Central Oregon	**OR - SIST**
Sunriver	Central Oregon	**OR - SUNR**
Tillamook	Oregon Coast	**OR - TILL**
Troutdale, McMenamins	Northern Oregon	**OR - TROU**
Wallowa Recreation Area	NE Oregon	**OR - WALL**

Washington

Friday Harbor	Island in NW Washington	**WA - FRID**
Leavenworth	Central Washington	**WA - LEAV**
Lopez Island	Island in NW Washington	**WA - LOPE**
Orcas Island	Island in NW Washington	**WA - ORCA**
Packwood	Central Washington	**WA - PACK**
Pasayten Wilderness	Northen Washington	**WA - PASA**
Ranger Creek	Central Washington	**WA - RANG**
Roche Harbor	Island in NW Washington	**WA - ROCH**
Tieton	Central Washington	**WA - TIET**

Wyoming

Alpine	Western Wyoming	**WY - ALPI**
Boyer & Savery Ranches	Southern Wyoming	**WY - BOYE**
Pinedale	Western Wyoming	**WY - PINE**
Saratoga	Southern Wyoming	**WY - SARA**
Thermopolis	Central Wyoming	**WY - THER**

Index   **Index**

The Equipment

Affectionately known as Romeo, N3790R has been used in researching all sites in the Flyer's Recreation Guides. The craft is typically loaded with one or two pilots and a full load of camping gear. Romeo's features follow:

- 1967 Cessna 172H
- 180 HP Lycoming, Avcon conversion
- Constant speed prop, Avcon conversion
- Horton STOL with gap seals
- Micro Aerodynamics vortex generators
- Auxiliary 18 gallon long range tank
- Cessna 310 front fork with large tire
- Large main tires

© 2001 by ALTA Research

Intro-9

Key

Boxes like this contain a quick synopsis of each chapter. The dollars at the end of the text indicate typical total cost for <u>two people</u> to stay overnight, eat meals, and enjoy entertainment. Example: $12-123 Info: 500-288-2582

Airport Information

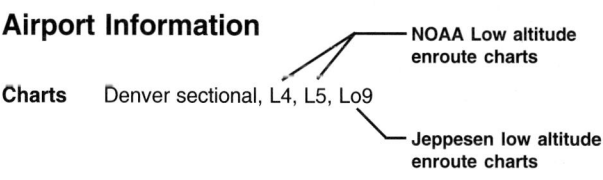

Charts   Denver sectional, L4, L5, Lo9

Lodging Information

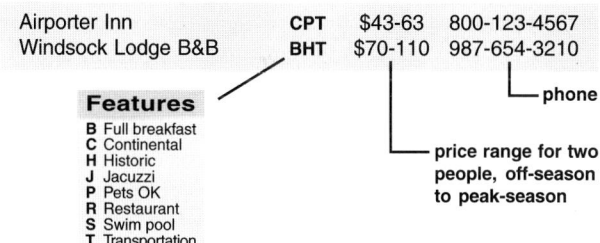

Things change! *Prices in the Guide include taxes and are based on 2001 rates. Phone numbers for all businesses are included – use them to verify current prices and availability.*

*See **http://www.alta-research.com** for update information.*

© 2001 by ALTA Research

DENSITY ALTITUDE CHART

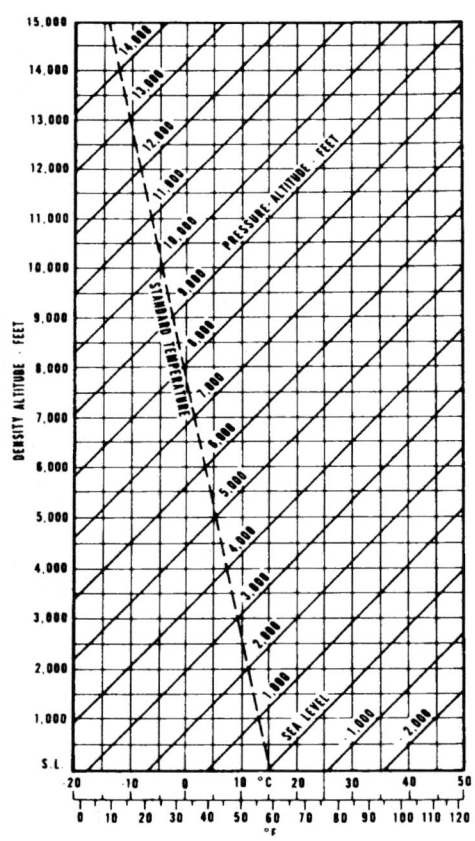

SOURCE: FEDERAL AVIATION ADMINISTRATION

Idaho

Idaho Aeronautics

208-334-8775

Idaho

© 2001 by ALTA Research

Big Creek

ID - BIGC

Big Creek Airport

Town	Big Creek **ID** U60 (U6-zero)
Coord	N-45-07.99, W-115-19.31
Elev	5743 feet, S end high
Runway	1rt - 19, 3550 x 110' turf
Freq	CTAF-122.9
Chart	Great Falls sectional

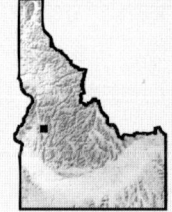

CAUTION: *Airport information not for navigational use. Mountains surround. Tricky drafts. Blind approach and departure. Announce intentions on 122.9. Runway has large dip at north end. Watch density altitude. Not for inexperienced mountain flyers.*

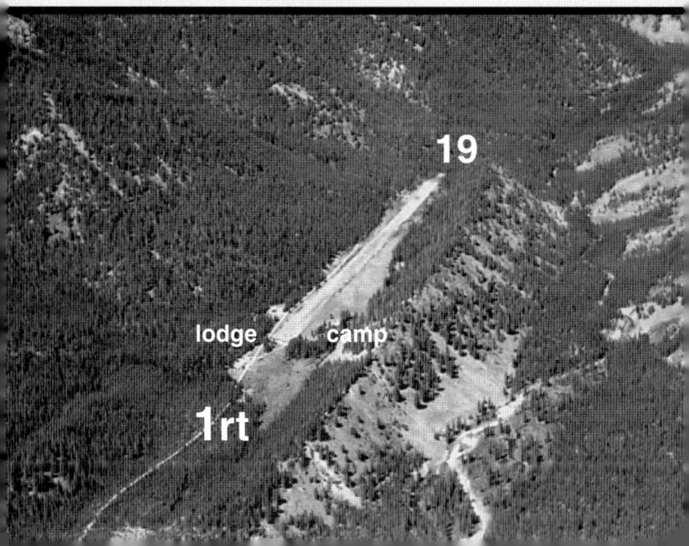

STATE OF IDAHO — BIG CREEK

AIRPORT LOCATION

LATITUDE: 45° 08' LONGITUDE: 115° 19'

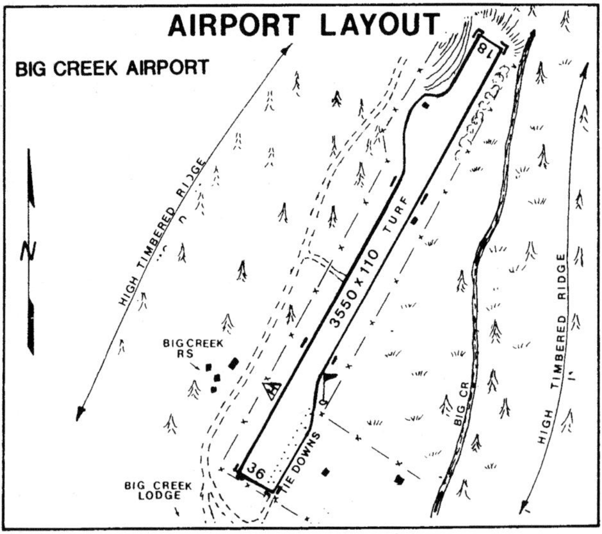

AIRPORT LAYOUT

BIG CREEK AIRPORT

Reprinted from Idaho Department of Transportation 1986 Airport Facilities Directory

© 2001 by ALTA Research

Big Creek **ID - BIGC**

> *Big Creek, a beautiful, challenging wilderness strip. Drop in for a hearty meal or stay and enjoy the wilderness. The facilities are rustic, but guaranteed to keep the bears out. Hiking, riding, fishing, and hunting. $0-100*
> *Arnold Aviation: 208-382-4336*
> *Idaho Aeronautics: 208-334-8775*

Airport Description

The airport is located at the western edge of the Frank Church River of No Return Wilderness, surrounded by peaks of over 9000 feet. It sits between a mountain to the west and a ridge to the east. The south end is somewhat higher than the north end. The surface is not maintained through winter months.

Land on Runway 19 unless conditions dictate otherwise. The strip contains a low point about 1000 feet from the north end, which may be somewhat muddy – consider landing on the last two thirds of the runway.

Depart on Runway 1. When departing, bear right, following the river to the northeast until you reach a safe altitude. Due to the blind departure route, announce departure on 122.9 and take serious care to avoid following an upstream tributary.

Services and Lodging

No phone or fuel are available. However, the nearby lodge and Forest Service have backcountry radio service. The U.S. Forest Service (USFS) can contact fire towers for local weather conditions. FAA weather reports are difficult to get.

Several tiedowns with strong chains are available. Except during hunting season, good tiedown space is normally available. Pleasant, **no-fee campsites** are located conveniently close to the tiedown area.

Big Creek Lodge is located less than 1000 feet south of the airstrip. In 1984, the proprietors expressed plans for expanding their sales to groceries and other improvements. Except for the a hydroelectric generator and higher prices, the lodge hasn't changed much since then. In 1996, the owner passed away. New owners have purchased the lodge. Verify the following before you visit.

Meals at the lodge are generous. The lodge keeper will cook you a hearty breakfast for $9, lunch for $7, and dinner for $15. Rooms costs $50 and cabins are $100. Call 888-848-0011 (not at site) for information and reservations. Advance reservations are recommended during hunting season.

Gillihan's Lodge is a B&B, located one mile south. Try contacting them on 122.9 for pickup. They have been known to serve lunch to drop-ins. Other meals are served to overnight guests, only. Reservations are required, 208-327-0907 (not at site). In my experience, communication with Gillihan's has been spotty.

You can pass messages to Big Creek or Gillihan's Lodge via backcountry radio by calling Arnold Aviation, 208-382-4336.

Transportation

Conventional transportation is unavailable at Big Creek. The airport is located at the end of the road. Fortunately, many trailheads can be reached by foot or bike. The lodge had loaner bikes when I visited. Better yet, ask the wrangler at the Big Creek Lodge for a trusty steed. Or, a lodge hand can truck you to distant locations by vehicle for a reasonable fee.

© 2001 by **ALTA** Research

Big Creek   **ID - BIGC**

What to Do

Big Creek Lodge is a small, rustic mountain lodge that enjoys most of its activity during **hunting** season; lodging is more likely to be available in summer. The proprietors can arrange **horse-camping excursions** during the summer and early fall. During winter, the mountain roads are closed. However, air taxi services can deliver people to Idaho lodges for **cross-country skiing** or other winter pursuits.

Outfitter Brian Simi provides guided **trail rides**, **pack trips**, **fishing excursions**, and **hunting trips**, 208-382-4872 (not at site). A two hour ride costs $35, for example. A seven-day hunting adventure costs $3200 to $4200 per person.

If you bring a **folding bike** or like to **hike**, you can double your options for adventure. You can visit **historic mines**: Werdenhoff Mine, Independence Mine, Red Metals Mine – or just enjoy the **incredible scenery**. Excursions range from sev-

Big Creek Lodge

eral to twenty miles. Galen Hanselman's <u>Fly Idaho!</u> describes a number of interesting places to visit.

For a pure **wilderness experience**, you can **hike** your choice of two- and three-day loops from Big Creek. Contact the Krassel Ranger District for further information, 208-634-0600. In particular, ask for a copy of their *East Half of Payette National Forest Recreation D-6* map, $7. This map shows all USFS-maintained trails for the area. The USGS topo maps for the area are *Big Creek* (north) and *Edwardsburg* (south).

Camping permits are not required. However, it pays to check in with the USFS located adjacent to the strip. The rangers are a good source of free maps and valuable verbal information.

The Big Creek area is rightly termed "backcountry" – sparsely inhabited and very rugged. The four-season residents are indeed a tough breed. Though I noted the occasional wearing of side-arms, the residents are friendlier than the norm for our "civi-

Packing into the wilderness

lized" cities. What little civilization there is, ends at the airstrip and lodge; one can easily walk from the airstrip to wilderness terrain. Scenic day hikes from the Big Creek area are available, and one such hike is described below.

Lick Lake / Ridge Hike

This hike can be compressed into a day-hike, or expanded for several days of backpacking into the wilderness. You pass Lick Lake, which is cold, but not so cold as to prevent a refreshing dip after an aggressive climb. Further on, you are treated to nice views from the ridge.

From the airstrip, walk Road 340 south for about 35 minutes. Look for the trailhead sign, *Lick Lake*. Turn left on the trailhead and follow Trail 065 east across Big Creek. Because there is no bridge across this creek, you must wade across. Continue following the trail.

The official entry to the River of No Return Wilderness is located about ten minutes from the trailhead. Just past the marker, drinking water is available from a small creek that gushes down the mountain. Local Forest Service personnel claim the mountain water is safe to drink. The trail crosses other sources of water at comfortable intervals.

The trail to Lick Lake is about two miles in length, and involves an elevation gain of almost 2000 feet. I clocked two hours from trailhead to Lick Lake. The trail is good, and scenic beauty lies ahead. As you near the Lick Lake checkpoint, watch for a sign that announces its position to the right. The lake is not far from the main trail, and suitable campsites are nearby.

For a rewarding panoramic view of the area, continue another hour on the trail to the ridge. From the ridge, you can continue to Bear Lake or follow other trails into the wilderness area. For further exploration, a topo map is advised.

- RW

© 2001 by ALTA Research

ID - BRUC

Bruce Meadows

Bruce Meadows Airport

Town	ref. Stanley **ID** U63
Coord	N-44-24.93, W-115-19.01
Elev	6370 feet
Runway	5 - 23, 5000 x 140' turf
Freq	CTAF-122.9
Chart	Salt Lake City sectional

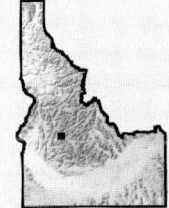

> ***CAUTION:*** *Information not for navigational use. SW end can be soft and wet. Post at Runway 5, and pole at 23. Animals on runway.*

STANLEY U63 BRUCE MEADOWS

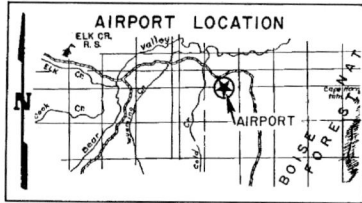

LAT 44° 24.9' N
LONG 115° 18.9' W

VOR	FREQ	RAD	NM
DNJ	116.2	099°	44 NM

FSS BOISE (BOI) 1-800-632-6582

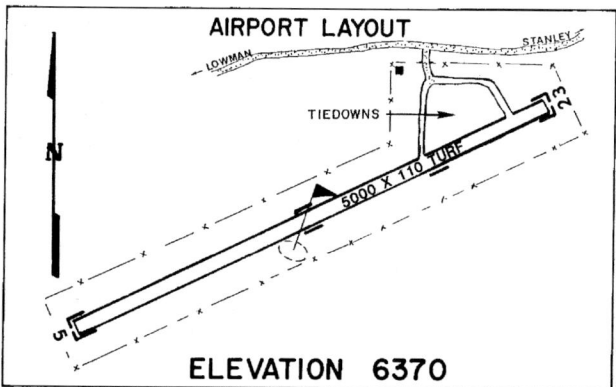

ELEVATION 6370

LOCATION: 20 MILES NW OF STANLEY **FUEL:** NO
LIGHTS: NO **COMMUNICATION:** CTAF: 122.9
FSS: 121.1T, 116.2R
REMARKS:

RECOMMEND LAND RWY 5, TAKE OFF RWY 23 WHEN WIND CONDITIONS PERMIT. AIRPORT IS LOCATED IN A HIGH MOUNTAIN VALLEY SURROUNDED BY HIGH TERRAIN. VERY HIGH DENSITY ALTITUDES IN SUMMER. AIRCRAFT TIEDOWN AREA IS ROUGH. CLOSE FLIGHT PLAN PRIOR TO LANDING. NO WINTER MAINTENANCE.

Reprinted from Idaho Department of Transportation 1990 Airport Facilities Directory

Bruce Meadows **ID - BRUC**

> ***Bruce Meadows, a relatively easy mountain airstrip with camping, wilderness hikes, and trails that lead to secluded hot springs. $0***
> ***Ranger: 208-259-3361***

Airport Description

Unlike most wilderness strips in the Idaho backcountry, the Bruce Meadows airport sits in a relatively wide, flat valley - plenty of maneuvering space between the mountains. A dirt road passes by the airport close to the tiedown area. The windsock is at mid-field. Aircraft typically land on Runway 5 and depart on 23.

Most mountain pilots would view this turf landing strip as wide and adequately long. The surface is a mix of gravel and sod with grass. It has been relatively smooth for my craft, but has the potential to be noticeably rough for airplanes with small wheels. Especially in spring, take care to avoid potentially wet and soft surface at the southwest end.

Airport Services, Lodging, and Transportation

Airport information: Idaho Aeronautics, 208-334-8775 (not at airport). Bruce Meadows is a backcountry airport without fuel, phone, lodging, or transportation. The tiedown area has a pit toilet, picnic table, fire pit, and shade cover for campers. Other places to camp are noted below.

© 2001 by ALTA Research

Bear Valley Hot Springs

The Bear Valley Hot Springs are a series of unspoiled wilderness hot springs and cascading pools beside the roaring Bear Valley Creek. This is one of the few last best places that has not had its soul removed by government or commercial meddling. Access to the springs can be a tad challenging, which is likely why the site remains pristine and untainted.

The hot springs are a six mile hike from the airport. They are located upstream from where Bear Valley Creek joins Marsh Creek, the headwaters for the Middle Fork of the Salmon. There are two routes that eventually parallel Bear Valley Creek. The first follows the river on the north shore, and then crosses to the south shore. The second is accessed via the Fir Creek Camp-

Bear Valley Creek is most easily crossed after mid-summer

ground, from which it follows the south shore until joining the other trail. Option two avoids the river crossing, but passes along dangerous cliffs above the river. I believe the latter is too risky when carrying a heavy pack.

There is no easy choice here. The river can be too high for crossing in spring, and the cliff route is risky any time of year. The best plan is to wait until after late July, and take option one.

From the tiedown area, follow the dirt/gravel road east for 20 minutes until the intersection with signs: *Fir Creek Camp Ground, Marsh Creek Trail, Blue Bunch Trail.* Turn left, and the campground entrance road is another 5 minutes on your right. The campground is a good place to stay if you do not want to carry a backpack to the hot springs.

Happy camper reads a book by the hot pools

Continue walking straight another 5 minutes, past the campground road, and to the Bear Valley Creek bridge. Here, you can self-register for the hike. Total time to the bridge is 0:30.

Cross the footbridge, turn right and follow the trail downstream. In spring and early summer, rafters float by. After 30 minutes (1:00 total) you arrive at the river-crossing point. My GPS indicated N-44-25.705, W-115-16.255. Because the rocks are slippery and the current can be deceptively strong, I use a walking stick when fording the river.

Continue following the trail downstream. The path crosses three minor hillside creeks. Expect to arrive close to the hot springs at about 2:10. The first clues will be a tree with "HS" carved into the bark and a partial clearing on the left toward the river. Here, you may notice a light trail toward the river and steam from the springs.

Continue a little farther for better accesses. The most desirable soaking pools and campsites are at the far end. In this general area, my GPS read N-44-26.648, W-115-14.324, and 3.9 nm from the airport.

Various hot pools cascade all the way into Bear Valley Creek. Excellent unimproved shaded campsites are close by. There are no toilets, picnic tables, or official fire rings. Depending upon time of day and year, the insects can be mildly annoying. Apply

Bruce Meadows **ID - BRUC**

some insect repellant to tweak reality, and the scene couldn't be more perfect – roaring river, hot water, interesting topography, and trees that reach to the sky. Evidence of wildlife abounds, and quiet visitors will likely see deer or something larger.

Protocol at the hot springs is clothing optional – some do and some don't. Pool temperatures ranges from too hot, to cool by the river. Sulfur smell at the lower springs is minimal.

Bruce Meadows is just a short hop from the Sulfur Creek Ranch **ID-SULF**, a good stop for breakfast, lunch, or dinner.

- RW

The warm riverside pool is perfect on a sunny afternoon

© 2001 by ALTA Research

ID - CHAM

Chamberlain

Chamberlain USFS Airport

Town	Chamberlain **ID** U79
Coord	N-45-22.74, W-115-11.81
Elev	5765 feet
Runway	7 - 25, 4100 x 200' turf, rough
	15 - 33, 2700 x 140' turf, bad
Freq	CTAF-122.9
Chart	Great Falls sectional

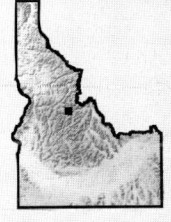

CAUTION: *Airport information not for navigational use. Watch density altitude; early morning and late afternoon use recommended in summer. Turbulence, sink. Rising terrain to N and W. Can be closed for mud in spring. Training activity. Moose.*

CHAMBERLAIN BASIN — CHAMBERLAIN

U.S.F.S. AIRPORT LOCATION

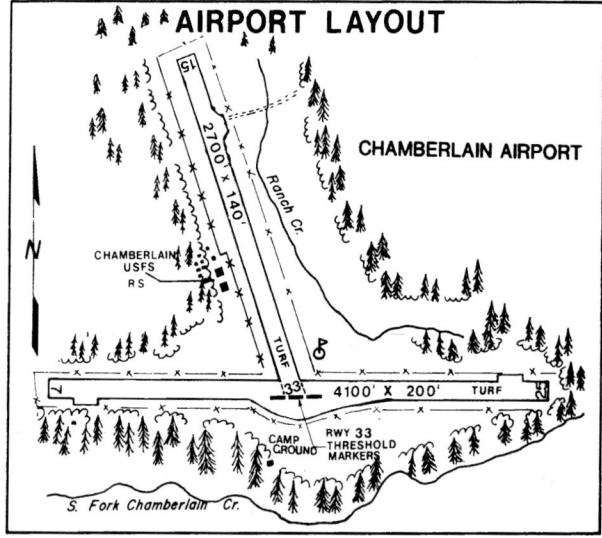

LATITUDE: 45° 23′ LONGITUDE: 115° 12′

AIRPORT LAYOUT

Reprinted from Idaho Department of Transportation 1986 Airport Facilities Directory

Chamberlain Basin **ID - CHAM**

> ***Chamberlain, a mountain flyer's airstrip, set deep within the River of No Return Wilderness. Great for primitive camping, hiking, and fishing. $0***

Airport Description

Before departing for Chamberlain, call the U.S. Forest Service (USFS) at 208-634-0600 for status, and ask local air taxi people about current runway condition. Idaho Aeronautics is another good source of information, 208-334-8775.

Chamberlain is located in the middle-northern area of the River of No Return Wilderness. Although not sandwiched between steep mountains like most Idaho wilderness airports, don't let the higher terrain to the north and west sneak up on you.

Runway 25 is the favored runway for arrivals. For noise abatement, avoid overflying the USFS cabins adjacent to Runway 15-33. The USFS requests no touch-and-goes. Conditions permitting, Runway 7 is favored for departure. Have your departure route clearly in mind before takeoff. Announce intentions on 122.9 before landing or departure.

Services, Lodging, and Transportation

Phone: USFS, 208-634-0600 (not at airport). Services and lodging are limited to information from the Forest Service and a nearby campground. This is a wilderness airport – no roads come within miles of the airstrip.

© 2001 by ALTA Research

Overview

The River of No Return Wilderness encompasses all of the old Idaho Primitive Area and Salmon River Breaks Primitive Area, plus additional wild lands totaling 2,361,767 acres. It is the largest national forest wilderness in the lower 48 states. A network of 296 trails totaling 2,616 miles link the various airfields, rivers, and trailheads; 116 bridges enable the trails to cross the rivers.

Positioned in the middle of a designated wilderness area, Chamberlain Basin has become a major drop-point for fishermen and hunters. Access is not possible by land vehicles. The USFS likes to keep motor noise to a minimum. Other than airplanes that come and go, motorized vehicles are not allowed. The runway is maintained by hand labor and mule power.

This is a good site for those who prefer **easy-to-moderate hiking**. The trails begin at a high elevation, so you can taste the mountain air without having to work hard for it. The trails are adequately maintained and well marked. In the past, bears had taken to eating the trail signs (made of California redwood), but the USFS claims to have licked the problem by switching to a less tasty flavor of wood.

Chamberlain Creek, which runs parallel to Runway 7-25, contains small **trout** that can be caught without too much skill or luck. Salmon are protected. **Campsites** are positioned between the runway and the creek, and also near the trees east of 15-33.

Contact the Krassel Ranger District for further information, 208-634-0600. Consider purchasing a copy of their *Payette National Forest Recreation D-6* map. They sell a waterproof,

Chamberlain Basin

ID - CHAM

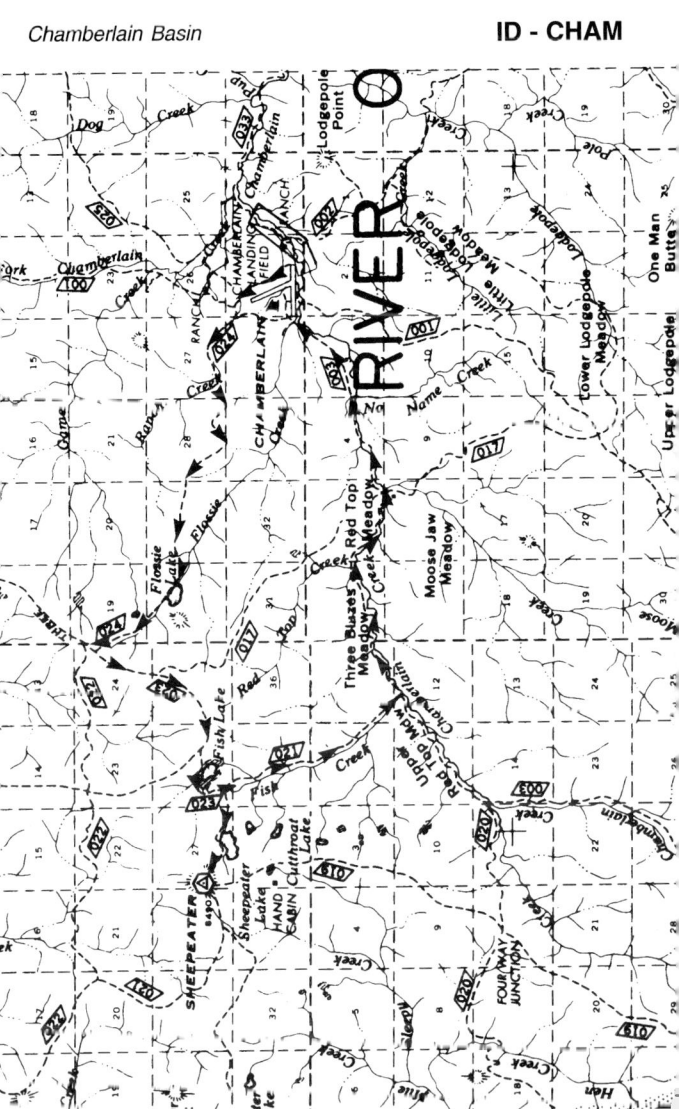

two-part colored version for $7 per half. These maps show all USFS-maintained trails for the River of No Return Wilderness.

Neither camping nor hiking permits are required, but check in with the Forest Service next to Runway 15-33. The USFS is a good source of maps, information, and drinking water.

Hiking options are abundant and charted. The **No Name Creek area** is beautiful, but trails are often light or nonexistent. The **Fish Lake Loop**, on the otherhand, is 22 miles of trails with a total elevation gain of several thousand feet. The average hiker can **backpack** the loop in three days. Trails are usable June to November, with a good chance of snow in November. If you need better detail than provided in the U.S. Forest Service maps, purchase the following USGS topos: *Lodge Pole Creek, Mosquito Peak, Sheepeater Mountain*, and *Meadow of Doubt*.

The Fish Lake Loop begins at the southwest end of the airport. Head southwest on Fish Creek Trail (#003 and #021) toward Red Top Meadow and Fish Lake. When you get to Fish Lake, consider walking past Sheepeater Lake up to the nearby Sheepeater Lookout. From Fish Lake, loop clockwise on Three Blaze Trail (#023) and Flossie Trail (#024) to Flossie Lake. Continue following Flossie trail back to the Chamberlain airstrip. Of course, this loop works in either direction.

Fishing and Hunting

Formal **hunting** and **fishing trips** are offered by Mackey Bar Outfitters, 800-854-9904. Hunting trips cost $4200 per person for seven days.

Trout fishing is viable in Chamberlain Creek, which runs parallel to runway 7-25. Check regulations, as not all species are in season.

- RW

© 2001 by ALTA Research

ID - DIXT

Town of Dixie

Dixie Town Private Airport

Town	Dixie **ID** none
Coord	N-45-33, W-115-27.5
Elev	5835 feet, N end high
Runway	17 - 35, 3000 x 80' dirt/grav
Freq	CTAF-122.9
Chart	Great Falls sectional

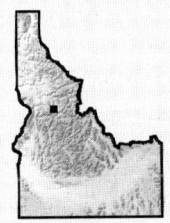

CAUTION: *Information not for navigational use. One-way, marginal go-around strip. Land N; depart S. Surface crooked and wavey. Mud.*

Dixie is a small mountain town, far from civilization. A starting point for pack trips, hikes, or just a place to get away. $0-250
Info: 208-842-2523, 208-842-2417

Airport Description

Two nearby Idaho backcountry airports have the name "Dixie." Do not confuse Dixie Town with the Dixie USFS airstrip 3.5 miles southwest, **ID-DIXU**. The private Dixie Town airstrip is not on the charts. Just a block from town, its meandering runway lies parallel to the Dixie's main street. The town is west of the runway, and terrain rises sharply to its east.

The runway is dirt and gravel. The north end is high. The south portion has a low spot, but the extreme south end is higher

Short final for Runway 35

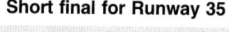

than the low spot. Expect mud patches after rain. Land north and depart south. Tiedown is at midfield, approximately across from the footbridge. When departing, follow the river until you have adequate airspeed to pull away.

Services, Food, and Lodging

Airport phone: 208-842-2523 (Lodgepole Pine Inn). There is no fuel, phone, or cell phone service at the strip, but there is a pay phone outside the general store. Camping is tolerated at the airstrip. If you want more privacy, just walk a short distance in almost any direction and camp on Forest Service land.

To get to town, walk across the bridge west, then one block to the main street. Turn left and walk another block to the Lodgepole Pine store, restaurant, bar, and inn. Silver Spur Outfitters and Lodge is close to the airport, about 150 feet from the windsock. Both offer three meals per day and lodging.

Old mining cabin

Town of Dixie **ID - DIXT**

The owners of the Lodgepole Inn manage the store, restaurant, bar, and inn, 208-842-2523. Rooms typically cost $42. The facilities are aesthetically plain, but quite adequate. The photo to the left shows the view from one of the rooms.

Silver Spur Outfitters and Lodge (previously Dixie Outfitters) has recently been remodeled and expanded. Owners Rick and Debbie Koesel can be reached at 208-842-2417, or on backcountry radio frequency 4634.5 KHz. They provide room and three meals for $65 per person. Multi-day pack trips average $100-125. Day rides are also an option. Afterwards, they can drive you to Red River Hot Springs to soothe the muscles.

Dixie

Dixie was founded on August 28, 1862 by two miners who trekked over the divide from Elk City and discovered gold in Dixie Gulch. By 1900, Dixie had become a boom town. During World War I, Dixie became nearly deserted as residents joined the service. By the depression, people were moving back to eke out a living by mining a little gold.

In 1964, the airstrip was built by Bob Black and others. Electrical power arrived in early 1980. Dixie remains highly isolated from civilization – over 80 miles of mountain road to Grangeville, the nearest town of significant size. The nearest town of insignificant size is Elk City, nearly 30 miles north.

Follow any trail into the woods, and you will stumble on **historic artifacts** such as old log cabins and abandoned mines. Consider carrying a GPS on such adventures as insurance against disorientation. Walk south on the main road 3/4 mile to visit Dixie's **old cemetery**. The oldest stone in this picturesque plot reads 1893. A campground with pit toilets is located near the cemetery.

Dixie USFS airstrip (see **ID-DIXU**) is farther south. It provides access to the Salmon River somewhat below the strip. Back in town, reasonably good **fishing** can be found in the stream by the airport, which is stocked with fish.

Most town activity centers around the store, restaurant, and bar. One way or another, residents are not afraid to speak their mind. I noted the following signs. The signs went up after Earth First attempted to curtail area logging.

© 2001 by ALTA Research

Town of Dixie **ID - DIXT**

> **Official Clinton-Free Zone**
>
> **Help destroy the USA, join Earth First**
>
> **We reserve the right to refuse service to all Earth Firsters and associates**
>
> **This public telephone is on private property. Earth First!ers using the telephone will be limited to only one person on this porch at one time!**

As most Idaho towns, the 4th of July is a big deal in Dixie – serious drinking, music, laughter, and fireworks. One group of retired folks comes from all over the U.S. to get their kicks by cutting lumber at the old outdoor mill just north of the runway.

Dixie USFS Guard Station

ID - DIXU

Dixie USFS Airport

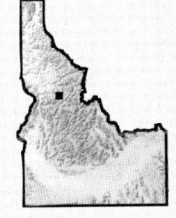

Town	Dixie **ID** ID05
Coord	N-45-30.84, W-115-30.77
Elev	5148 feet, N end 56' high
Runway	18 - 36, 4788 x 120' turf
Freq	CTAF-122.9
Chart	Great Falls sectional

CAUTION: Information not for navigational use. One-way, marginal go-around strip. Land N; depart S. South end soft, and tall grass is possible.

DIXIE

DIXIE USFS

LAT 45° 31.3' N (30.84)
LONG 115° 31.0' W (30.77)

VOR FREQ RAD NM

DNJ   116.2   014°   55

FSS BOISE (BOI) 1-800-632-6582

AIRPORT LAYOUT

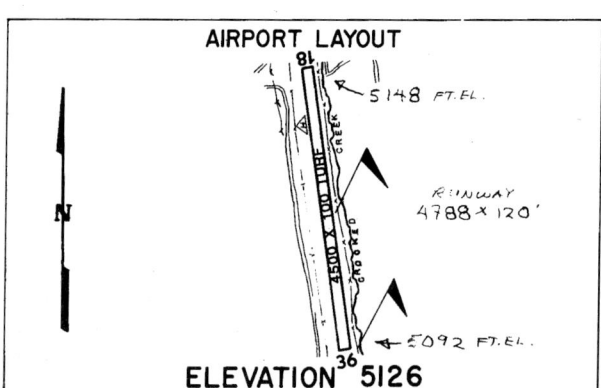

5148 FT.EL.

RUNWAY 4788 x 120'

5092 FT.EL.

ELEVATION 5126

LOCATION: 3 MILES SW OF TOWN   **FUEL:** NO

LIGHTS: NO   **COMMUNICATION:** CTAF: 122.9
                                                       FSS: 121.1T, 116.2R

REMARKS:

RECOMMEND LAND RWY 36, DEPART RWY 18, RIGHT TURN DOWN CROOKED CREEK, EXTENSIVE HELICOPTER ACTIVITY IN SUMMER MONTHS. CLOSE FLIGHT PLAN PRIOR TO LANDING. NO WINTER MAINTENANCE.

Reprinted from Idaho Department of Transportation 1990 Airport Facilities Directory

Dixie USFS Guard Station **ID - DIXU**

Dixie USFS Guard Station is an end-of-the-road airstrip with convenient campsites and trailheads that lead into the Idaho wilderness. $0
Info: 208-842-2237, 208-842-2255

Airport

The runway is in a valley, surrounded by mountain ridges. Note that the Dixie Town airstrip (**ID-DIXT**) is roughly 3.5 miles northeast, a possible source of confusion. Landing to the north is recommended because it provides an up-hill roll, and because the south end is more likely to have a soft surface. When landing, roll full north before exiting the runway. This proce-

dure will keep you on firm ground and out of drainage ditches at either side of the runway.

When approaching directly from the south, you must drop quickly after the southern ridge in order to make the runway. Be prepared to bleed altitude at a rapid rate. A last-minute go-around can be risky, especially for marginally powered aircraft. This is the approach I used when landing at the strip.

Alternatively, consider arriving via the narrow canyon from the west. While this approach avoids the rapid descent of the former, it requires a quick turn to the left followed by quick alignment just before landing.

When viewed from the ground, the departure route may be far from obvious. The most practical departure in most cases is south. I concluded that the safest procedure for my aircraft was to achieve a safe margin of airspeed for a turn, then quickly turn

Camper prepares a meal near the tiedown area

Dixie USFS Guard Station **ID - DIXU**

west and follow the narrow canyon as it winds its way to lower elevations.

Tall grass, soft surface, and ground squirrel holes are potential problems. The USFS typically begins cutting grass and filling holes in early July.

Services, Transportation, and Lodging

Dixie USFS phone: 208-842-2237, Elk City USFS phone: 842-2255. The rangers are not aviation experts, but they can provide useful information about runway surface. The ranger station at the northeast corner of the runway is a good source of local information. They have a phone that they may let you use, but no cell phone service or aviation services are available. Their UHF radio frequencies are 411.475 and 445.475.

The northwest corner has a tiedown area and several nice campsites, one of which is by a creek. A pit toilet and garbage cans are provided. Firewood is available, but it may be unsplit.

Halfway House is a more secluded campground, located in the canyon that exits the valley. It is literally at the end of a long series of roads that penetrate the Idaho backcountry, with trailheads that enter the wilderness. One trail takes you 12 miles to the Salmon River. Halfway House has it all: nice stream, tall trees, picnic tables, pit toilet, fire wood, and improved fire rings. To get there, follow the dirt road near the tiedown area south for 1.3 miles.

Both of these no-fee, no-bureaucracy campgrounds are a credit to the local rangers.

Dixie Town is about 4 miles northeast by car. If you decide you do not like the looks of the Dixie Town airstrip, you can use the Dixie USFS airport as an alternate. Dixie inn keepers can usually provide transportation.

© 2001 by ALTA Research

Dixie USFS Guard Station

The Dixie Guard Station came to life between 1924 and 1934 with the construction of a saddle and feed room, forest ranger's log house, and cook house. In 1930, the Forest Service began building the Dixie Air Field.

The road system, built by the Forest Service in 1938, allowed placer miners to take advantage of the higher gold prices of the 1930s. The renewed gold mining created yet another boom for the area. The Forest Service was also responsible for installing telephone lines to the outside world.

The Guard Station is presently used as a work center and fire camp during high fire seasons. The airstrip is still used in fire-fighting operations.

- *RW*

Signs at the guard station tell visitors where to go

Flying B Ranch

ID - FLYB

Flying B Airport

Town	none **ID** none
Coord	N-44-58, W-114-44 est.
Elev	3647 feet
Runway	16 - 34, 2000 x 50' turf
Freq	CTAF-122.9
Charts	Great Falls sectional

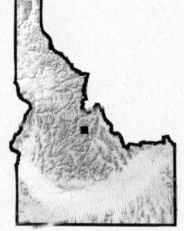

CAUTION: *Airport information not for navigational use. Short strip for altitude. Restricted maneuverability. Watch for traffic from Bernard airstrip, north. Use 122.9. Private; landing permission advised.*

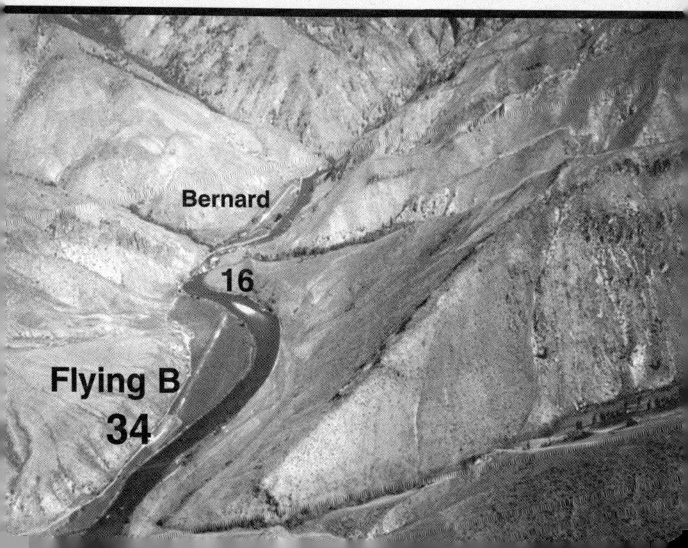

Bernard

16

Flying B
34

2   *Idaho*

Flying B Ranch **ID - FLYB**

*The Flying B, a guest ranch in the heart of the lower 48's most rugged, isolated mountain range. Many miles from the nearest road or phone, a challenging mountain strip. Fish, hunt, ride, hike, or just enjoy the view. $150-800
Info: 208-756-6295, flyingrr@ida.net*

Airport Description

Like a good number of the Idaho wilderness airports, the Flying B strip sits below towering mountains at the edge of a roaring river – in this case, the Middle Fork of the Salmon River. It is located less than a mile south of the public Bernard Creek Airstrip. At first glance, maneuvering space, visibility, and go-around capability seem to favor landing on Runway 34. However, local professional pilots go either way.

Considering the elevation, short runway length, and limitations in maneuverability, this strip is best left to experienced mountain pilots. The Flying B is a private airstrip; land with prior permission only. Have a meal at the lodge or stay the night, and you may be spared the $25.00 landing fee. Announce your arrival on 122.9. Like magic, a jeep arrives before you cut your engine. If the magic doesn't work, it's a pleasant 15-minute walk.

Services

Airport phone: 208-756-6295 (not at airport). Other than a few tiedowns, the strip has no fuel or services for aircraft. Backcountry radio and satellite phone are the only high tech com-links with the outside world.

© 2001 by ALTA Research

Flying B Ranch

The Flying B is located at the southeastern edge of the Frank Church Wilderness. It is one of the few wilderness guest ranches that operates year round. This is no mean feat; the nearest road and utilities are past a mountain range some 30 miles to the east by air, and several times this distance by foot. The ranch communicates to the outside by single sideband HF backcountry radio and has its own hydroelectric power plant that runs year round. The strip is plowed, allowing planes with suitable wheels to land any time of year. (Verify in advance!)

This narrow stretch of flat land was home for a settlement of Sheepeater Indians until the late 1800s when Captain Bernard arrived, followed by a few hearty settlers. The site became a marginally viable cattle ranch in 1912, and eventually was sold to A. A. Bennett during WW-II. Bennett, a WW-I fighter pilot and Ford Trimotor bush pilot, was well known in these parts. He had laid out a number of air strips in the Idaho back country prior to developing the Flying B fly-in resort.

Flying B Ranch **ID - FLYB**

Today, the ranch is operated by the Flying Resort Ranches, Inc., an organization of 150 members who support the ranch in various ways (mostly funding). In return, members pay lower prices than guests. The ranch is operated by a paid staff of 7 to 10. When I visited, I was graciously treated to lunch by two members from Montana. Guests and staff are friendly and relaxed.

The ranch is tucked close to the river, with mountains rising on either side. The main lodge is the focal point, surrounded by a corral, semiprivate simulated log cabins, and utility buildings. An impressive little suspension bridge crosses the river. You wonder, how did that get here? On the other side, trails lead past a hot-hot spring, and off to adventure. It's a pleasant layout.

The minimum cost for each nonmember is $105 per day. This includes three hearty meals and lodging. A deposit is required in advance. Camping on the premises is not an option. Fly-in visitors can purchase individual meals: breakfast for $14, lunch for $14, and dinner for $18. The meals are "hearty," not to be confused with gourmet.

The ranch doubles as a stop for all rafters from outfitters like the Mackay Bar Corporation, 208-344-1881. Rafters arrive at the river bank, stretch their legs, take a siesta on the grass, or purchase film and snacks from the Flying B commissary. A clever arrangement of fences and gates gives rafters access to the store, without allowing them full run of the grounds.

The prime activities at the ranch are **day rides**, **pack trips**, **fishing**, and **hunting**. In between, you **hike**, **shoot skeet**, eat, get to know other escapees, and sleep like a log. Fishing is mostly catch and release, though one can legally bag selected fish when in season. Expect to catch salmon, steelhead, bull trout, and others. Skeet shooting costs $17 for 25 shells.

Pack trips on horseback run a minimum of three days. The cost is $135 per person per day. They offer a popular trip that takes you along a 67-mile trail, northwest and up to their high-country Root Ranch at 5600 feet. This second facility has a small lodge, four bunkhouse-style cabins, and a cook shack. It also has a 2100-foot private airstrip, which requires a good mountain aircraft and skills to match.

Like most of the wilderness operations, **hunting** season is the most exciting time of year. Lodge-based hunting has a four-day minimum. Spike camping excursions are six days at $431 per day per person. This, no doubt, sounds rather excessive to the non-hunter. Consider that a good deal of effort goes into setting up remote wilderness hunting camps, which are used a small portion of the year. For the serious, financially secure sportsman, the hunt of a lifetime may well be worth $2600 or more. Game includes deer, elk, moose, goat, sheep, mountain lion, and bear. You can hire a day-guide for bear hunting at about $125 per day. Mountain lion hunting occurs in January and February. Burrrr!

For details, write the Flying Resort Ranches, Inc., #9 Hamner Drive, Salmon, ID 83467; or call 208-756-6295.

- *RW*

Johnson Creek

ID - JOHN

Johnson Creek Airport

Town	Yellow Pine **ID** 3U2
Coord	N-44-54.73, W-115-29.14
Elev	4933 feet, S end high
Runway	17 - 35, 3400 x 150' turf
Freq	CTAF-122.9
Chart	Great Falls sectional

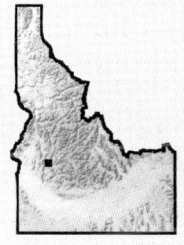

CAUTION: Airport information not for navigational use. Airport is located deep in narrow valley; hazardous when windy. Watch for irrigation pipe on runway. Turf extends takeoff roll. Watch density altitude.

STATE OF IDAHO — JOHNSON CREEK

AIRPORT LOCATION

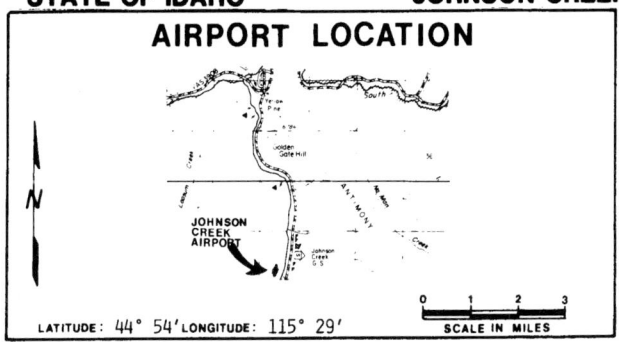

LATITUDE: 44° 54' LONGITUDE: 115° 29'

AIRPORT LAYOUT

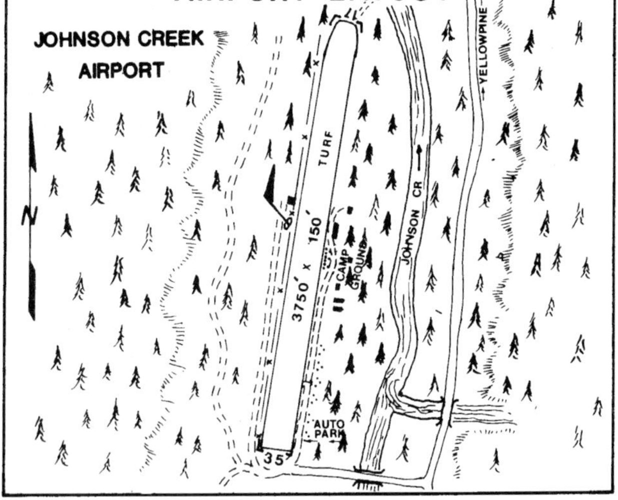

Reprinted from Idaho Department of Transportation 1986 Airport Facilities Directory

© 2001 by ALTA Research

Johnson Creek **ID - JOHN**

Johnson Creek is a peaceful mountain strip with camping, fishing, mountain hot tub, an isolated mountain town, and up-scale resort. $0-800 Lynn Imel: 208-633-4635, imel@ruralnetwork.net

Airport Description

Johnson Creek airport is located adjacent to and west of Johnson Creek. It is surrounded by mountains that rise 4000 feet on either side of the airport. The airport can be circled, but circling should be done with caution at pattern altitude. The north end of the runway is 40 feet lower than the south end.

Because the surface climbs to the south, landing on Runway 17 and departure on 35 are recommended. I know of at least one unfortunate pilot who departed south and was forced to set down in Johnson Creek. Yellow Pine, three miles north, is 300 feet lower than the strip. Circle above Yellow Pine to gain elevation.

Services, Lodging, and Transportation

Phone: Idaho Aeronautics, 208-334-8775 (not at airport). Idaho Aeronautics has provided sturdy tiedowns. A pilot's pay phone has been recently installed. The nearest source of food is a bar with grill in Yellow Pine, about three miles north by road.

Camping facilities and rest rooms with flush toilets are located between the airstrip and Johnson Creek. A **horseshoe pit** is provided for airport guests. Firewood is free for the asking. Bunk houses are available for those who are forced down due to bad weather. All of these amenities are provided courtesy of the state of Idaho.

© 2001 by ALTA Research

In Yellow Pine, Yellow Pine Lodge (a.k.a. Rosenbaum's) has a half-dozen rooms, 208-633-3377. Rooms cost $32-50, breakfast is $4-8, and family-style dinners are $8-13. For reservations or pickup from the airport, call 633-3377. The Yellow Pine Store rents a couple rooms behind the store and a number of Spartan cabins for $20-40. They pickup from the airport, and will even deliver groceries to your airplane. Contact the store at *ypstore@ruralnetwork.net* or 633-3300.

At the other end of the scale and five miles south, the Wapiti Meadow Ranch offers a wilderness-country retreat with gourmet meals, tasteful environment, and class excursions.

Years ago, the Idaho Aeronautics Board provided two inexpensive courtesy cars until an intoxicated customer destroyed one of the vehicles and then proceeded to sue the state. The cars were removed, and for years pilots had to depend on their creativity to get around. A shuttle service is now available from airport caretakers.

Overview

The airport is located on the southwestern edge of the Frank Church River of No Return Wilderness. As with most mountain airports, use this airport with caution. The strip itself is bright green turf, meticulously irrigated and maintained. The nearby airport camping facility is one of the best-kept airplane camping facilities in the West, and is large enough to handle fly-ins.

For a taste of old-time mountain culture, visit Yellow Pine, the small **mining town** three miles north. Weekends are lively, featuring occasional bar fights as entertainment.

A festive black powder shoot occurs in February. On the first full weekend in August, the town sponsors a three-day **harmonica festival**. For information, contact Lynn Imel, 208-633-4635.

© 2001 by ALTA Research

Johnson Creek **ID - JOHN**

Lynn and husband Dave often monitor 122.9, and they are eager to help pilots with backcountry needs.

Hiking trails that access the west end of the Idaho Primitive Area can be found east of the airport. The USGS *Yellow Pine* topo map covers this area. **Fishing** is available several hundred feet from the airport.

Wapiti Meadow Ranch

Five miles south of the airstrip sits the Wapiti Meadow Ranch, an upscale guest ranch for those who value quality and are willing to pay for it. Besides comfortable lodging and **gourmet meals**, the ranch offers access to virtually every kind of **moun-**

Yellow Pine streets come alive during the annual Harmonica Festival

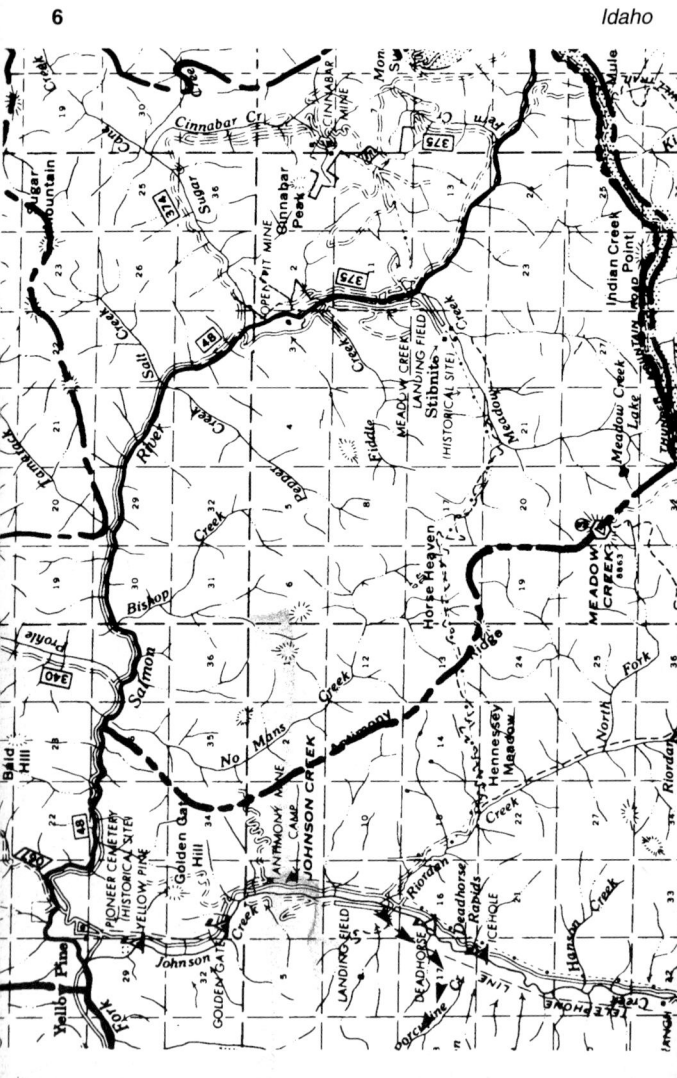

Johnson Creek **ID - JOHN**

tain entertainment available. If they cannot meet your special need themselves, they know an outfitter who can.

Historic artifacts oddly suggest that ancestors of the Incas, Mayans, and Aztecs once inhabited the valley. The Nez Perce and Shoshone tribes moved in, and were eventually succeeded by rugged miners, trappers, and homesteaders in this still primitive land.

Clark Cox established the first guest ranch at the site in the early 1920s. After operating for five decades, the Cox ranch eventually passed to Diana and Barry Bryant. Sharing complementary skills in entertaining, backcountry recreation, and mountain flying, they have transformed the ranch into the ultimate wilderness retreat.

The log and stone lodge is surrounded by cabins where guests sleep and grab moments of privacy. Guest cabins have multiple rooms heated by wood and electricity, stocked with the snacks of your choice. Pets are not allowed. Meals and snacks are decid-

edly gourmet. I'll spare you the details – just reading their sample menu makes my mouth water. If you catch some fish, the chef will prepare your prize to perfection.

Fly fishing is a speciality of the ranch, and they are an Orvis Endorsed Fly-Fishing Lodge. Guiding, complementary equipment, and new gear are available at the ranch. Fish the ranch's **trout pond** for Rainbows; or let the guides take you to local streams for Cutthroat, Rainbow, and Dollies.

Other summertime activities include **hiking**, **horseback riding**, **pack trips**, **4x4 excursions**, **visiting mines**, **panning for gold**, **mountain biking**, soaking in the **outdoor spa**, evening **campfires**, and **photography**. In winter, there's **cross country skiing**, **snowmobiling**, **snowshoeing**, and cozy moments by the hearth. The airport is used through winter; skis advised. Exercise due caution.

Summer prices range from $280 to $500 per person per day, including tax and 15% service charge. Packages are typically 3- or 6-days and generally include everything. Off-season prices for bed and meals alone are $100-200 per person; shorter stays can be accommodated. For terms and details, contact the owners at 208-633-3217, or email *wapitimr@aol.com* . The web site is *www.wapiti.com* . On arrival, guests may call on 122.9 for pickup.

© 2001 by ALTA Research

Johnson Creek **ID - JOHN**

Although I have not had the opportunity to visit the ranch myself, recommendations from others justify my passing the word to readers. Diana Bryant kindly provided the photos.

Mountain Hot Springs

A hot tub fed by **hot springs** is perched on a mountain side within a 30-minute walk, southwest of the airport.

Walk to the trailhead north of the windsock at the west edge of the runway. Look for the sign shown to the right of the windsock. From the sign, follow yellow markers west and then south along a trail that traverses the edge of the Bryant Ranch. Because a good portion of the trail crosses the Bryants' private property, please honor their wish that you remain on the trail. Continue following the trail south at the base of the mountain ridge.

After about 20 minutes, the trail swings west and begins to climb. The terrain becomes open and rocky as you climb toward the tub. The trail becomes more difficult to follow, so keep a sharp eye for yellow markers, and eventually, the white bathtub.

Other than the white tub and black hoses, the area is in its natural state. The tub's location provides a good view of the valley and surrounding mountains. The black hoses can be used to fill the tub. When I visited, the water temperature was slightly warmer than lukewarm. Just lean back and enjoy the mountain view.

- RW

Mountainside wilderness hot tub, a 30-minute walk from the airport

ID - LAVA

Lava Hot Springs

Lava Hot Springs Airpark

Town	Lava Hot Springs **ID** ID26
Coord	N-42-36.5, W-112-01.9
Elev	5268 feet, S end high
Runway	14 - 32, 3500 x 100' turf/dirt
Freq	CTAF 122.9
Charts	Salt Lake sectional

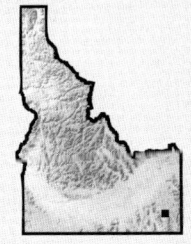

CAUTION: Airport information not for navigational use. Mtns nearby. Crosswinds common. Check for animals on runway. Surface muddy after rain. Airport not attended. Call Tom Lawler at 208-776-9745 for surface condition.

> *Lava Hot Springs, a small country town with a pleasurable resource – millions of gallons of mineral-rich hot springs water, offering a variety of options for soaking and swimming. Skiing and snowmobiling in winter. $55-140*
> *www.ci.lava-hot-springs.id.us*

Airport Description

Lava Hot Springs Airpark sits on a long mesa, 200 feet above the valley, surrounded by mountains, one mile southwest of Lava Hot Springs. Pilots have commented that the topology and runway orientation make the airpark look like an aircraft carrier.

The runway is marked at intervals with white cones. Red windsocks are located at either end. Quartering crosswinds are not uncommon. Most pilots land uphill on Runway 14, and depart downhill on Runway 32.

Although the airport is generally usable all summer, it is not attended; hazards may be present. Possibilities include: livestock, long grass, mud, and rodent holes. When you taxi, watch out for electric fence wire. Six robust chain tiedowns are available at the southwest edge of the runway. Look for black aviation-tire markers.

Call Tom Lawler in Lava for a condition report before landing, 208-776-9745. Tom is a commercial pilot who can provide you with recent information. Although the airstrip is restricted, it is my airport, and you have permission to land at your own risk. It is not insured, so take care, and have an enjoyable stay.

Lava Hot Springs **ID - LAVA**

Airport Services and Transportation

Phones: Tom Lawler at Lava, 208-776-9745; Reed White at Alta Research, 541-929-5738. Fuel: none. At time of publishing, there were no services at the airpark. There are no public phones, but cell phones work great. Given advance notice, inn keepers are usually willing and able to pickup from the airport, so consider giving them a call before you depart.

Pocatello is the nearest full-feature airport. Phones: FBO, 208-234-6154; and manager, 232-7440. Fuel: 100LL and jet. The Pocatello AvCenter has three courtesy cars. Contact the AvCenter at 234-2141 or 122.95. Enterprise provides car rental service to the airport; the others have offices at the airport:

Avis	888-897-8448	208-232-3244
Enterprise	800RentACar	208-232-1444
Hertz/Overland West	800-654-3131	208-233-2970
Budget Rent-a-Car	800-527-0700	208-233-0600

To get to Lava Hot Springs from Pocatello Airport, head east on Highway 86 toward Pocatello, then Highway 15 toward Salt Lake City. After 25 minutes on Highway 15, exit east on Highway 30 and drive another ten minutes to Lava Hot Springs.

Lodging and Dining

Lodging in Lava Hot Springs is both diverse and inexpensive. You may camp at the airpark, but there are no amenities. Camping is available on the Portneuf River at Ranch Inn, 208-776-9917, for $8 per tent, and in town at River's Edge, 776-5209 for $15. You can camp for $10 at the northwest corner of the Lava Hot Springs Airpark at Smith's Trout Haven, 776-5348. Inns follow:

© 2001 by **ALTA** Research

Dempsey Creek Lodge	P	$40-70	208-776-5000
Home Motel	HT	$30-90	208-776-5507
Lava Hot Springs Inn	BHJPT	$55-100	208-776-5830
Lava Spa Hotel	JT	$50-90	208-776-5589
Oregon Trail Motel	PT	$40-100	208-776-5050
Riverside Hotel	CHJPT	$55-115	800-733-5504
			208-776-5504
Royal Hotel	CHJT	$79-120	208-776-5216
Tumbling Waters Hotel	T	$65	208-776-5507

Although most inns will pickup from the Lava Hot Springs Airpark, verify and make arrangements in advance. Some inns will give you a discount if you mention this book. All inns mentioned here are within easy walking distance of everything in town.

Features

B Full breakfast
C Continental
H Historic
J Jacuzzi
P Pets OK
R Restaurant
S Swim pool
T Transportation

The hot springs water at all facilities in Lava is non-sulfur, and safe enough for drinking in some cases. Three historic inns

Lava Hot Springs **ID - LAVA**

have **hot springs soaking facilities**: The Home Hotel, Lava Hot Springs Inn, and Riverside Hotel.

The historic Home Motel has an outside spa, 26 inexpensive rooms, and a large house that rents for around $80. Most rooms have large tubs fed by hot spring water. The Lava Hot Springs Inn has outside bathing. The Riverside Hotel has been nicely restored with tasteful decor and private soaking tubs.

The Lava Spa and Tumbling Waters Hotels are located less than one block from the Portneuf river and the Idaho Foundation's outdoor hot springs baths. These motels provide the conventional amenities in every room: air conditioning, TV, and phone. The Oregon Trail Hotel offers a variety of room styles. The Royal Hotel is a four unit B&B with Italian restaurant.

My favorite place for dinner is a wonderful little Thai restaurant located by the river at the west end of town, just west of the bridge. The Riverside serves great Chinese meals, weekends only. I also enjoy Italian meals at the Royal Hotel and breakfast at Johnnys, both on Main Street. The Shed restaurant will pickup from the airport for a barbecue dinner, 208-776-9420.

Evening entertainment options range from a peaceful soak in the hot pools to carousing at Lava's two hot spots: the Blue Moon and the Wagon Wheel. One or both provide **rock** or **country music** on weekends, with **karoke** filling the gaps.

Lava Hot Springs Recreation

Lava Hot Springs is a rustic, friendly village nestled by the Portneuf River between the Wasatch Range and western flank of the Rockies. As its name suggests, Lava has been blessed with a generous flow of hot mineral water.

The various inns and B&Bs have a variety of hot water systems with which they tempt their guests: tubs in each unit, indoor tubs, outdoor tubs, wooden tubs, showers, and pools.

© 2001 by **ALTA** Research

Besides the private establishments, Lava has two exceptional public facilities at either end of town. You can't miss the modern state-owned spring-fed **warm swimming pool** at the west end of town. This unusually shaped pool is huge, with Olympic lanes, a multiplicity of diving platforms, and a new slide. It is open in summers, only.

Fortunately, the state's **hot mineral pool** complex at the east end of town is open year round except Thanksgiving and Christmas. For $5.50 entry or $6.00 day pass, the aesthetically designed facility offers mineral rich soaking at up to 110 degrees F without a trace of sulfur odor. The pools are refreshed with 3,500,000 gallons of spring water daily. Call 800-423-8597 or 208-776-5221 for more information.

An **unimproved hot spring** is located very close to town. Known locally as the Chicken Soup Spring, this spring dumps

Heated swimming pool with high-dive and water slide

Lava Hot Springs **ID - LAVA**

into a circle of rocks at the edge of the Portneuf River. To get to the springs, locate the blue Idaho sign on East Main to the right of the Lava Spa Motel. Walk behind the motel and follow the trail east along the river. You arrive at the springs within five minutes. Within 20 feet of the neglected soaking area you will see an opening in the river bank that looks like a mine shaft. Some have used this cave as a form of sweat lodge.

Although the main draw of Lava Hot Springs is the water for which the town was named, Lava provides other leisurely diversions that work well between soaks. The Idaho sign, which says "Idaho Centennial Trail," is a trailhead for **hiking trails** that head south and up the mountain. Or, rent an **inner tube** and float down the Portneuf River. Tube rentals are available just steps from the mineral pools. At times, a shuttle "train" pulled by a disguised John Deere tractor provides shuttle service for tubbers. The town's **Tube Festival** is in August.

Idaho Foundation Hot Springs

Play a round of **golf** at Lava's 9-hole golf course, 1.5 miles south of town by the airpark. Or, simply take a stroll around town and visit the local **history museum**. Surrounding areas have won a reputation among sportsmen for **good fishing** and **trophy-quality deer**, **elk**, and **moose**.

The Portneuf River below Lava is open year-round for **fishing**. Rainbow trout, cut throat, and German browns, all ranging up to ten pounds, await the fly or spin fisherman. Pebble Creek, Topance Creek, and Dempsey Creek yield delicious trout. Fish without a license at **Smith's Trout Haven**, 208-776-5348, southwest of town near the airpark: $0.25 per inch, or $2.50 per pound after one pound. During winter, local folks ice fish at the Chesterfield reservoir.

The Baker Ranch, 208-776-5684, offers a family **horse-drawn hay ride** through Lava's scenic backcountry. Their 4.5 hour ride

Tubers can float through town on the Portneuf River

includes entertainment and a **trail dinner** that won't leave you hungry. Adults cost $18; children $12.

Fish Creek Ranch, 208-776-5055 or 251-6316, provides **trail rides** to the top of Mt. Moh, locally known as the "L Mountain." They also have a free **petting zoo** and **horse shoe pits**.

Somerville Manor operates the **Medieval Living History Farm**, open to visitors in summer. Entry is $3.50; children free. Check for events at 208-776-5429 or *www.somervillemanor.org*

The surrounding mountains offer good **deer hunting**, and trophy bucks are not unusual. Black bear can be shot during the regular season. The Portneuf River offers good **duck hunting**, mostly mallards. The Chesterfield Reservoir area is a good spot for Canada and snow geese. In the mountains, grouse hunters have a choice of blue grouse, spruce grouse, and sharptail.

Winter snows bring new options: **cross-country skiing**, **downhill skiing**, and **snowmobiling**. The **Pebble Creek Ski Area** is located 20 miles northwest of town near Inkom. Although runs are available for all levels of skill, Pebble Creek is known as one of the most challenging ski areas in the United States. The facility features 2000 vertical feet of skiing on Mt. Bonneville, includes great alpine skiing, and offers guided backcountry tours. Rentals, ski packages, lessons, and day care are available. Call 208-775-4452 for details.

Snowmobilers enjoy miles of meticulously groomed trails by day, to be rewarded by a hot soak in the springs at night. Challenging trails climb the likes of nine thousand foot Mt. Bonneville. Cross-country trails connect Pocatello, Idaho Falls, and Alpine, Wyoming.

Local events at Lava include **Pioneer Days** in July and the **River-Rafter Tube Festival** in August. Smith's Trout Haven, 208-776-5348, sponsors a **Mountain Man Rendezvous** on the weekend closest to July 24th.

© 2001 by ALTA Research

Lava Hot Springs Airpark

I purchased the land for the Lava Hot Springs Airpark in 1993 – a wild idea that came from sitting in a hot pool too long. Being a lover of hot springs and most anything that flies, I decided that Lava would be the perfect site for a recreational and residential airpark. My objective for the project:

> *To work within an aviation community to help create the most interesting small airport within a radius of 500 miles.*

The turf runway was built to FAA guidelines in 1993. Word spread, and like the Field of Dreams, visitors arrived. As it grows, the airpark will become home for pilots and aviation-related businesses.

To assure that the development maintains its "aviation-first" priority, covenants require that residents have aviation interests, and that businesses be aviation related. Future plans include an aviation bookstore, aviation fuel, FBO, maintenance service, restaurant, and motel. Please get in touch if you would like to start a business like one of the ones listed.

The airpark is located on a 200-foot high mesa, providing a view of surrounding mountains and valley for every residence and business. Covenants assure that all buildings and landscape will be of top quality, and harmonious with the environment.

If you are like me and have always dreamed of living and working in a community like this, the Lava airpark may be the place for you. Ten residential lots from 1.5 to 6.8 acres are plated. Commercial opportunities are wide open. Call Reed White at 541-929-5738 or 877-360-ALTA. I look forward to answering your questions and hearing your ideas. For information about additional nearby properties to the east and west of the airpark, call 208-776-5536 and 307-883-4760, respectively.

© 2001 by ALTA Research

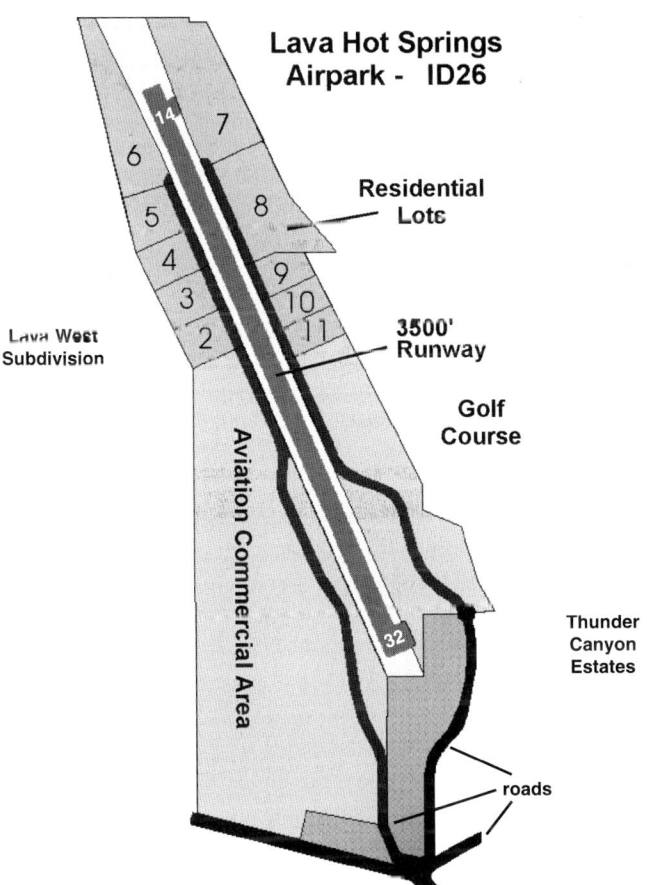

Pocatello Area

Home of Idaho State University, Pocatello and its 55,000 inhabitants serve as a regional center for higher education, cultural activities, and shopping. Three **golf courses** serve Pocatello.

The **Ross Park Zoo,** 208-234-6196, and **Fort Hall Replica**, 234-1795, are located southeast of town at 2900 South Second Street, close to Exit 67 from Highway 15. The Replica is inexpensive for adults, and even less for kids.

Fort Hall was coined "crossroads of the West" in 1835 when it served settlers traveling on the Old Oregon Trail. Although the original no longer stands, a trading post serves as a crossroads for modern-day travelers and Indian culture. The Shoshone-Bannock tribes offer **custom tours of the reservation**. Red Perry is the key contact for historical tours, 208-238-0097. For general information, call the administration at 238-3700, or the museum at 237-9791.

The Fort Hall Trading Post Complex is located north of Pocatello, off I-15, Exit 80. The annual **Shoshone-Bannock Indian Festival** is a yearly highlight in early August. Events include Indian games, sports, rodeo, parades, queen pageant, powwow, and Indian dances.

For evening entertainment, check the local papers for current theater and music concerts. The Green Triangle, 208-237-0354, is Idaho's largest nightclub, featuring six bars, antiques, dance lessons, contests, and live **country music** three nights a week.

Annual Pocatello events include the **Idaho Frontier Festival** in early August and **Fort Hall Indian Festival** (see above) in the week that follows. Call the Chamber of Commerce at 208-233-1525 for a complete list and specific dates.

© 2001 by ALTA Research — *RW*

ID - MOOS

Moose Creek

Moose Creek USFS Airport

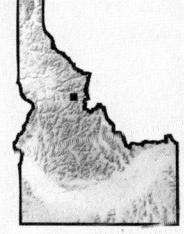

Town	Moose Creek **ID** 1U1
Coord	N-46-07.25, W-114-55.64
Elev	2454 feet
Runway	4 - 22, 2300 x 200' turf
	1 - 19, 4100 x 250' turf
Freq	CTAF-122.9
Chart	Great Falls sectional

CAUTION: *Airport information not for navigational use. Surrounding mountains. Trees. Tricky drafts. Blind approach and departure. Announce intentions on 122.9. Not for inexperienced mountain flyers. Call for conditions before landing.*

MOOSE CR. R.S.
MOOSE CREEK
USFS

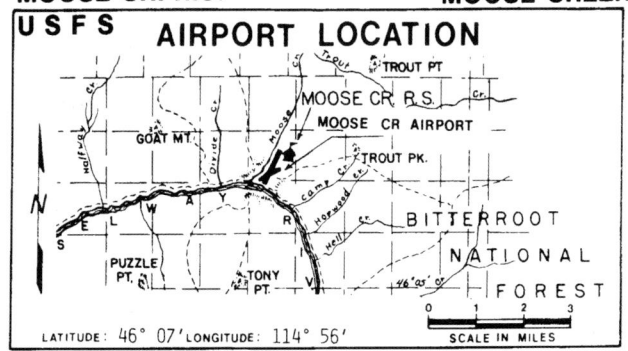

AIRPORT LOCATION

LATITUDE: 46° 07' LONGITUDE: 114° 56'

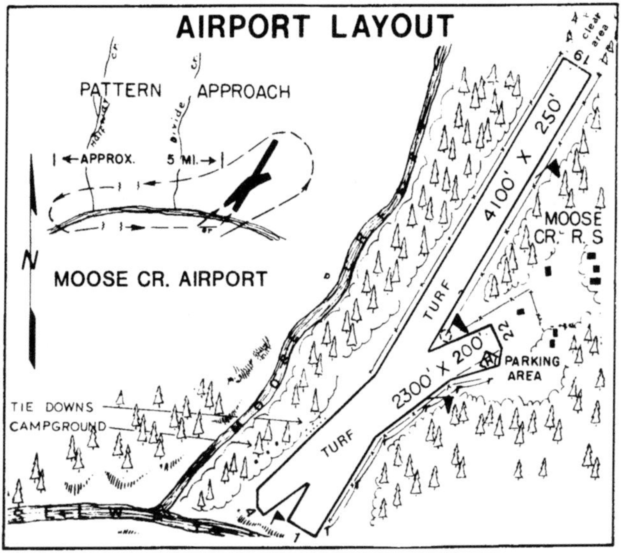

AIRPORT LAYOUT

MOOSE CR. AIRPORT

PATTERN APPROACH — APPROX. 5 MI.

4100' × 250'
2300' × 200'
TURF

MOOSE CR. R.S.
PARKING AREA
TIE DOWNS
CAMPGROUND

Reprinted from Idaho Department of Transportation 1986 Airport Facilities Directory

Moose Creek **ID - MOOS**

> ***Moose Creek is a USFS wilderness ranger station that serves as a hub for trails that radiate in all directions. Camp at a local campsite, or hike the Selway-Bitterroot Wilderness area. $0***

Airport Description

The airstrip is located on the north edge of the Selway River in the Selway-Bitterroot Wilderness in the Nez Perce National Forest. Mountains several thousand feet higher than the strip surround the narrow valley. Due to drafts, blind spots, and confining terrain, use extra caution. This airstrip is for experienced backcountry flyers only.

Runway conditions and availability vary with the seasons. In spring, runways may be too soft for landing. In summer, Moose Creek is sometimes closed as a fire-prevention measure.

The recommended approach is east along the Selway River for Runway 1 in summer and Runway 4 in early spring or late fall. Runway 4 has better drainage. Announce intentions on 122.9. A ranger in the Shissler Fire Tower sometimes helps control traffic. Watch for activities such as haying on the runways. Be prepared for downdrafts at the southwest end. The windsocks are difficult to spot because there are no circles around them. Expect a bumpy landing. The best tiedown area is at the northeast end of Runway 4, a short walk to the U.S. Forest Service (USFS) buildings.

Recommended departure Runway is 19 in summer and 22 if soil is damp. These runways slope down and have the best orientation for departure along the Selway river. Be certain that the weather is clear enough so that you will not run into low clouds over the river; there is no room to turn around. Winter operations are conducted with skis on Runways 4 and 22.

Before departing for Moose Creek, call Orofino Aviation at 208-476-4714 or Mountain Air at 983-9193 for current airport information. Verify runway conditions and procedures with a local air taxi pilot. For additional site information, contact John Foland, the Nez Perce Forest Air Officer, at 983-1964.

Services , Lodging, and Transportation

Phone: 208-476-4714 (not at airstrip). Except for information from the USFS, no services are available at Moose Creek. You might enjoy visiting the historical exhibit about the site.

Lodging is limited to well-separated campsites near the airstrip. The shady campsites and convenient tiedowns leave little to be desired. Just find a comfortable campsite just west of the runways. Facilities include grills, picnic tables, and rest rooms.

Because Moose Creek is in a designated wilderness, visitors must walk to see the sites. No roads or land-based motor vehicles can access the site. The nearest road is 22 miles away.

Permits

Backpacking and camping permits are not required, but the USFS requests that you fill out a card and deposit it in a box at the trailheads.

Fishing regulations allow selective fishing of the Western Slope Cutthroat. Due to endangered population balance, fishing regulations are enforced.

© 2001 by ALTA Research

Moose Creek **ID - MOOS**

Overview

Moose Creek may well be a USFS employee's paradise. The setting is very pleasant. Access is possible only by foot, hoof, or air. Food and supplies are packed in over a one-day trail by horse. At last check, Moose Creek was the only wilderness area with a non-flying motor vehicle. A tractor was flown in, piece by piece, years ago. The tractor is used for haying and maintaining the airfield.

Due to summer heat at lower elevations, the best time to visit is in June or early July. The field may be closed at other times for fire safety or other reasons, so remember to check in advance.

A number of well-marked trails radiate from the Moose Creek Ranger Station. If you would like to **hike** a loop by streams and through local mountains, plan on several days, minimum. Trails are steep, but well posted with destinations and mileages. The rivers are wide and often covered with whitewater. Crossings of the Selway are practical only with the provided bridges. Two such suspension bridges are located at the southwest end of the field.

For maps and more information, contact the Moose Creek Ranger District at 208-926-4258. Two maps cover the area. The airstrip is on the north half, but the south half is apt to be more useful. Ask for the large, colored *Selway Bitterroot Wilderness Maps* (waterproof: $8 each half) and the *Selway Bitterroot Primer* (Publication #R1-92-47, free). An excerpt from the Primer follows.

- RW

© 2001 by **ALTA** Research

THE RIDGERUNNER

He lived in constant fear, convinced all his life that someone was chasing him. For 24 years (1936 - 1960), the Ridgerunner fled through the mountains in and around the Selway-Bitterroot, pursued by his imaginary enemies. He hid from everyone, shot at low-flying planes, and always kept moving. Not even in winter and lack of snowshoes stopped his ridgerunning. For warmth, he wore a shirt made of a blanket and socks from dish towels. For food, he broke into backcountry cabins with a key made from the tin of a meat can braced with the broken blade of a jack-knife. To avoid capture, he seldom slept in cabins. His shelter even in blizzards was a torn piece of canvas.

Although few had ever seen him, there was no doubt he existed. His unique way of ruining every cabin he visited was his calling card. Opening cans and jars (he especially loved jam), he would eat a little from each container and leave the covers off so everything would spoil. Instead of removing the outside cap on the stovepipe, he would dismantle the pipe inside and smoke would fill the room, blackening the walls. He dirtied every dish and scattered garbage on the furniture and floor. His fame grew in Central Idaho.

In 1942, rumors spread that the Ridgerunner was Baldy Webber, a vicious criminal wanted for attempted murder. By raiding a trail crew camp, he finally pushed the Forest Service too far and they resolved to capture him.

© 2001 by **ALTA** Research

But even real enemies couldn't catch the Ridgerunner. For three years the Forest Service chased him through Selway country. He raided a trail maintenance camp on Rhoda Creek. Once when his pursuers were too close, he hid in a hollow tree on Roundtop Mountain. Finally, in February, 1945, after following his tracks in the snow for miles, the Clearwater National Forest's two best woodsmen, Morton Roark and Lee Horner, spotted smoke from his campfire. Knowing Webber to be dangerous, they armed themselves, separated, and moved towards the camp from two directions. Quietly, they crept closer. When they could almost touch him, the Ridgerunner glimpsed Roark's snowshoe. Roark ordered, "Don't Move!"

But huddled by the campfire was not Webber, the killer. The Ridgerunner was Bill Moreland, a tiny 5'2" man wearing rags and missing most of his teeth.

Nothing ever stopped Moreland — not even arrest. The judge liked him and after serving only 90 days in an Orofino jail, he rushed back to running ridges and messing up cabins. Eventually he acquired another bad habit. He shot at real people. Exasperated, the Forest Service recaptured him in 1958 by outrunning him with a helicopter. This time the courts committed him to a mental hospital in Orofino, but to no avail. He escaped in a year.

Only when he was too old to run did Moreland leave the mountains. In 1960, he voluntarily returned to the hospital retiring from ridgerunning forever.

© 2001 by ALTA Research

Murphy Hot Springs

ID - MURP

Murphy Hot Springs Airport

Town	Three Creek **ID** 3U0 (3U-zero)
Coord	N-42-01.24, W-115-20.30
Elev	5829 feet
Runway	1 - 19, 5250 x 120' turf
Freq	CTAF-122.9
Chart	Salt Lake City sectional

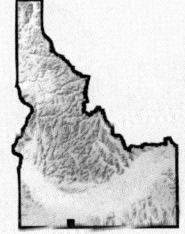

CAUTION: Airport information not for navigational use. Fences at ends. Tall grass. No line of sight between runway ends. Power lines cross canyon. Restricted area northwest.

MURPHY HOT SPRINGS ◯ MURPHY HOT SPRINGS

AIRPORT LOCATION

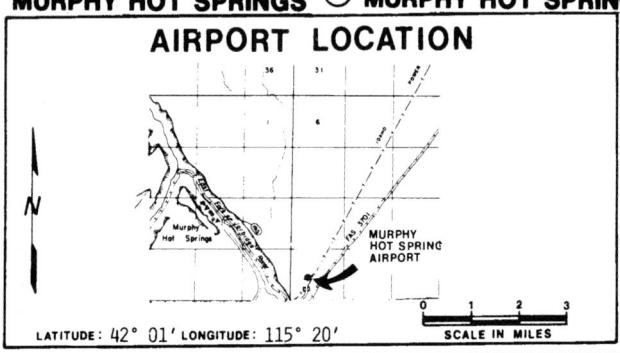

LATITUDE: 42° 01′ LONGITUDE: 115° 20′

AIRPORT LAYOUT

MURPHY HOT SPRINGS AIRPORT

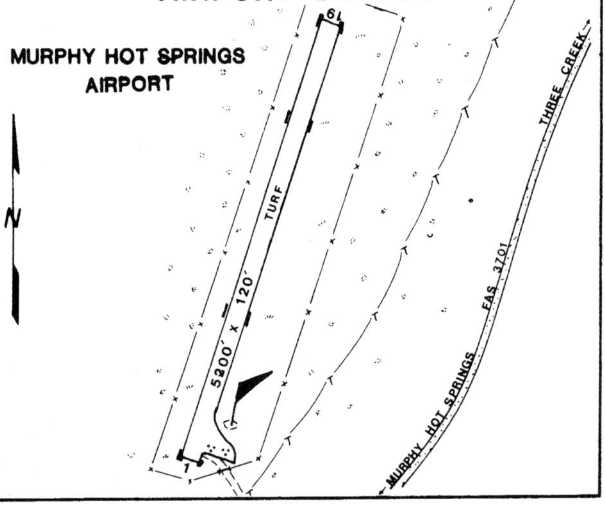

Reprinted from Idaho Department of Transportation 1986 Airport Facilities Directory

Murphy Hot Springs **ID - MURP**

> *Murphy a.k.a. Desert Hot Springs is a rustic mountain resort that jumped out of the 1920s. Low cost, non-alcoholic, and hot water to soak in.*
> *$6-70   Resort: 208-857-2233.*

Airport Description

The airport is an unembellished grass strip on a flat plateau, unattended with no lights. Look for a southwest-bound road and powerlines intersecting a northwest-bound canyon. The windsock is located near the tiedown area at the south end. Several tiedowns are provided. Although the resort's name has changed, airport guides may refer to the airport as Murphy Hot Springs.

Airport Services, Lodging, and Transportation

Phone: 208-334-8775 (not at airport). Call the local resort at 857-2233. No phones or services are located at the strip.

A car from Desert Hot Springs resort, the only nearby lodging, will pick you up for free if you stay the night. To catch their attention, the owners recommend a good, low buzz or two, but

prior arrangements are best. The resort is located in the canyon, two miles northwest of the airport. If you opt for the buzz, look for a pool next to the river. Avoid the powerlines that cross the canyon two miles southeast of the resort.

Desert (a.k.a. Murphy) Hot Springs

In wagon train days, the springs were called "Kitty's Hot Hole." When I first visited, owner Harry Showalter, had somewhat improved the facilities to have a 1920s flavor. His lodging consisted of nine small, primitive cabins. Matthew Olivas purchased the facility and changed the theme to a non-drinking environment with a natural bent. He also changed the name from Murphy Hot Springs to Desert Hot Springs. His cabins rent for $25 for a single, $35 for two beds, and $45 for three beds.

The resort includes an organic-style restaurant. The land is shaded by trees. A stream isolates the **hot springs** from the rest of the resort. You cross a footbridge to get to the hot springs. The hot springs include one large pool and private baths. The springs produce 129 degree water that is reduced to 100 degrees for the pool and limited to 115 degrees in the bath houses. The springs cost $4 per day.

Other area activities include **hiking**, **fishing**, **hunting**, **pack trips** with horses, and seeing the local scenery. For a fee, resort operators can drive you to places of interest, such as the old mining town of Jarbidge.

- *RW*

ID - SILV

Silverwood

Silverwood Airport

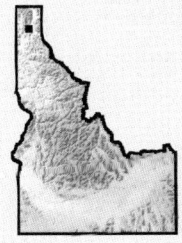

Town	Athol **ID** S62
Coord	N-47-54.49, W-116-42.56
Elev	2350 feet
Runway	3rt - 21, 4200 x 50 asphalt
Freq	CTAF-122.7, possibly 122.8
Charts	Seattle & Great Falls, L9, Lo5

CAUTION: Airport information not for navigational use. Trees near 21. Aerobatics, gliders, and air shows. No line of sight between runway ends. No snow removal.

> *Silverwood Theme Park, a modern amusement park with a variety of well-designed rides, games, things to see, and treats to eat. $75-150*
> *208-683-3400--0,   www.silverwood4fun.com*

Airport Description

The airport lies in open terrain with an amusement park close by its side. The windsock is mid-field northwest, but may be difficult to spot since its color blends with the ground. The airport is closed winters and at night.

In July and August, overnighters may wish to tiedown north of the museum to avoid fireworks debris.

Airport Services, Transportation, and Lodging

Airport phone: 208-683-3400, then dial 0. No fuel. Available are pilot supplies, rest rooms, and food services at the park. The airport is located adjacent to the theme park, so no transportation is needed to the park. Unfortunately, if you are simply stopping for lunch, you are still required to pay the $24 entry fee to access the restaurant.

The best deal in lodging is the Athol Motel, three miles north of the airport, 208-683-3476. Rooms cost $30-50. With advance notice, owner Elmer will pick you up at the airport.

The park has a campground with shade trees, bathhouse, and laundromat. To get to the campground, locate the tunnel entrance near the main admission gate. Follow the tunnel under Highway 95 to the camping area. Campsites cost $20.

© 2001 by ALTA Research

Silverwood Theme Park **ID - SILV**

Silverwood Theme Park

Silverwood Theme Park is a modern **amusement park** with well-designed rides, games, things to see, and treats to eat. All aspects of the park are integrated within manicured grounds. Though significantly smaller in scope than Disneyland, its quality is in the same league. The park opens at 11:00 daily in summer, and is open weekends only in spring and fall.

The entry fee may appear rather steep. However, consider that for $24, adults receive unlimited access to dozens of rides in the park. Children receive the same access for $12, and toddlers are free. The quality of the rides and **entertainment** is above the norm. The entertainment is live, and includes an Olympic-quality **ice skating show, fireworks,** and the outside possibly of an **air show**.

Silverwood Theme Park

The park's theme is turn-of-the-century Old West. A **steam locomotive**, **shops**, and **restaurants** all carry the theme, but the **carnival rides** bridge the gap from historic to modern day technology. The **Corkscrew Coaster** is a beautifully designed roller coaster that turns you over and upside down. The latest addition is the **Tremors Coaster**, an above- and below-ground roller coaster. **Flume rides**, **water bumpers**, **Ferris wheel**, **wooden roller coaster**, and an **antique carousel** are some of the other rides available.

Through 1994, the park included a micro zoo, featuring **Tigger the Tiger**, a lion, and horses which kids could ride. Now, Tigger enjoys status as the sole animal attraction.

Up until about 1995, the park had a strong aviation theme: a beautiful little aviation museum, antique planes, and daily air shows. Due to an unfortunate accident at another location, one of their top aerobatic performers was killed. Further, management made a business decision to sell the park's antique airplanes to finance the purchase of a wooden roller coaster.

Although the Norton Aero Museum is no longer a feature at Silverwood, the **air shows**, **biplane rides**, and **glider rides** may continue to be offered at some point in the future. For status and other information, call 208-683-3400. Because prices and offerings change at Silverwood, a preflight call is advised.

In spite of the disappointing loss of aviation attractions, Silverwood remains a quality amusement park. You'll get far more points for flying the family to Silverwood than for feeding them $100 hamburgers at the Dunbar Bar and Grill.

- RW

© 2001 by **ALTA** Research

ID - SMIL

Smiley Creek

Smiley Creek Airstrip

Town	Galena **ID** U87
Coord	N-43-54.91, W-114-47.84
Elev	7160 feet, N end lower
Runway	14 - 32, 4900 x 150 foot, turf
Freq	CTAF-122.9
Chart	Salt Lake City sectional

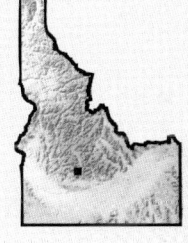

CAUTION: *Airport information not for navigational use. Fence, each end. Watch for irrigation sprinklers on runway. Watch density altitude.*

STATE OF IDAHO — SMILEY CREEK

AIRPORT LOCATION

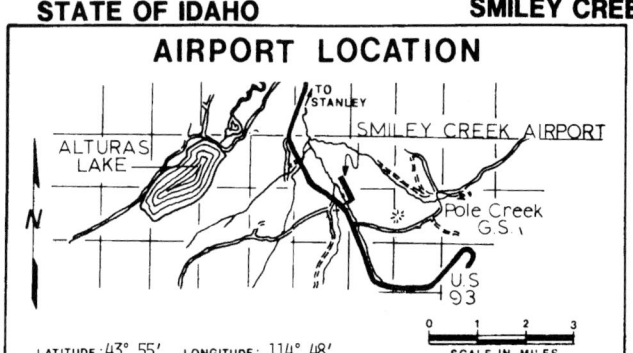

LATITUDE: 43° 55'   LONGITUDE: 114° 48'

AIRPORT LAYOUT

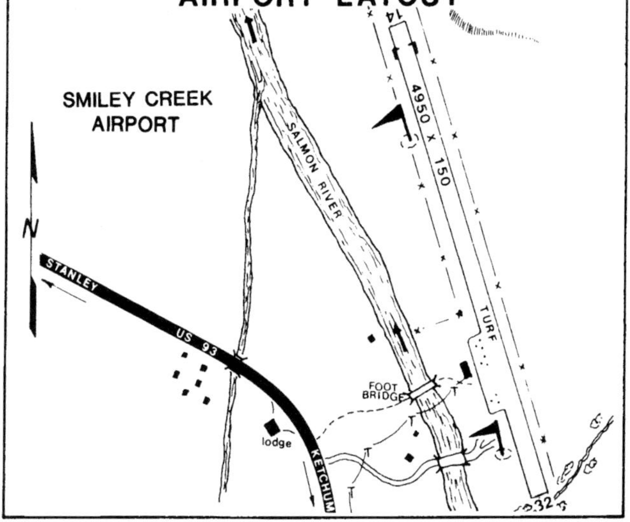

SMILEY CREEK AIRPORT

Reprinted from Idaho Department of Transportation 1986 Airport Facilities Directory

Smiley Creek **ID - SMIL**

> *Smiley Creek is a high-altitude destination with lodge at the strip and alpine hiking trails within a ten mile drive. $5-100*
> *Lodge: 208-774-3547.*

Airport Description

The Smiley Creek airport is a grass strip located within a wide valley east of the Sawtooth Mountains. Except for fencing at either end, operations at Smiley Creek are obstacle-free. Compared to most other mountain strips, Smiley Creek is relatively safe for landing. The turf conveniently reduces touchdown roll – a boon for landing, but a liability for takeoff. Because of this and the strip's elevation, consider departing when the air is cool if you are heavy or underpowered.

Wind permitting, land on Runway 14 and depart on 32.

Services, Lodging, and Transportation

Airport information phone: Idaho Aeronautics, 208 334-8775 (not at airport), or 774-2984. Fuel is not available. The closest fuel is 18 miles north at Stanley Airport. Two dozen well-secured tiedown chains are provided by the state of Idaho at the southwest edge. During summer, a caretaker keeps the field in tip-top shape. There are no lights and snow is not removed.

Smiley Creek Airport has a well-equipped campground with drinking water, clean restrooms with flush toilets, heated bathhouse with hot showers, shaded picnic tables, and grills. A rental car may be available from the caretaker. The car is provided by the state of Idaho for $5 per day plus $0.30 per mile.

© 2001 by ALTA Research

A rustic lodge with restaurant, bar, and grocery store is located not 500 yards from the airport. The grounds have been embellished with antique wagons and mining artifacts. The facilities include restaurant, gas station, store, and lodging.

If you plan to eat breakfast at the lodge, come hungry. Although breakfast seems to be the big draw, their other meals are equally satisfying – local trout and prime rib, for example. The chef even takes pride in his wine selections, a rarity in the backcountry, or in Idaho for that matter.

Lodging costs $58 for a 4-person cabin with wood stove, and $78-99 for lodge rooms. A tepee with electric blankets can be rented for $44. Meal prices are reasonable. Call 208-774-3547 for information or reservations.

Eleven miles north of the airport, you can stay at the historic Idaho Rocky Mountain Ranch. Their facilities were crafted of native lodgepole pine in 1930. They offer **horseback riding**, **fishing**, **horse shoes**, **mountain biking**, and soaking in their **hot springs pool**. The cost of $105-135 per person includes breakfast and dinner. Reservations are necessary, 208-774-3544.

See **ID-STAN** for more area lodging.

Overview

The white man first ventured into the valley in the early 1800s. However, the area remained home to Shoshone and Bannock Native Americans and a few hearty mountain men until the discovery of gold. In 1878, Levi Smiley discovered gold at the head of the creek, and the boom began.

The towns of Vienna and Sawtooth City sprang to life. Although Vienna quickly grew to a population of 1500 with some two hundred buildings, not much is left today. Today, the

Smiley Creek **ID - SMIL**

population center has gravitated twenty statute miles north to Stanley.

The airstrip is located five miles southeast of the Sawtooth Wilderness Area. A number of **hiking** options, all of them beautiful, are not far from the airport. You can ask at the lodge for tips on places to visit with scenery and old mines.

The Stanley Ranger Station at 208-774-3000 can provide maps in advance, and can also help you select a hike. They sell a waterproof *Sawtooth National Forest* map for $7, and also the *Alturas Lake* and *Snowyside* USGS topo maps. See **ID-STAN** for additional area information, including a list of outfitters.

Sawtooth Mountains

Alpine Creek Trail

This hike offers alpine scenic beauty, with only 2 hours penetration time and less than 2000 feet elevation gain. Bring the afore mentioned topo maps and some good mosquito repellent.

Transportation to the trailhead is quick and easy with the airport courtesy car. After leaving the airport, travel about three miles north on the main highway. Turn left on the road labeled *Alturas Lake*. Follow this road about six miles southwest. It passes Perkins Lake on the right and then Alturas Lake on the left. You pass several good car-camping sites on the shores of Alturas Lake. Watch for the Alpine Creek trailhead about one and a half miles after leaving Alturas Lake.

At the trailhead, take care to select the correct trail. Do not cross the wood bridge over the creek. Instead, look northwest from the parking lot to find the trailhead. A good clue is the sign that reads *Trail Not Usable By Motor Vehicles*.

For the first hour, the trail climbs gradually through a perfect mixture of sage, grass, alpine streams, and pine trees. You pass frequent open areas, revealing views of surrounding mountains. The first established campsite is located about 45 minutes up the trail near a stream. Depending upon time of year, mosquitoes can be a problem at this campsite.

At this point, the trail is still easy to follow. Trees are frequently engraved with trail markers. One hour from the trailhead, the trail begins a much steeper grade. My topo map (dated 1964) indicates no trail beyond this point. However, the trail continues its ascent, following the northeast branch of Alpine Creek in close proximity.

The trail enters the lake district at 8500 feet and becomes difficult to identify. The terrain is open enough to allow you to

Smiley Creek — ID - SMIL

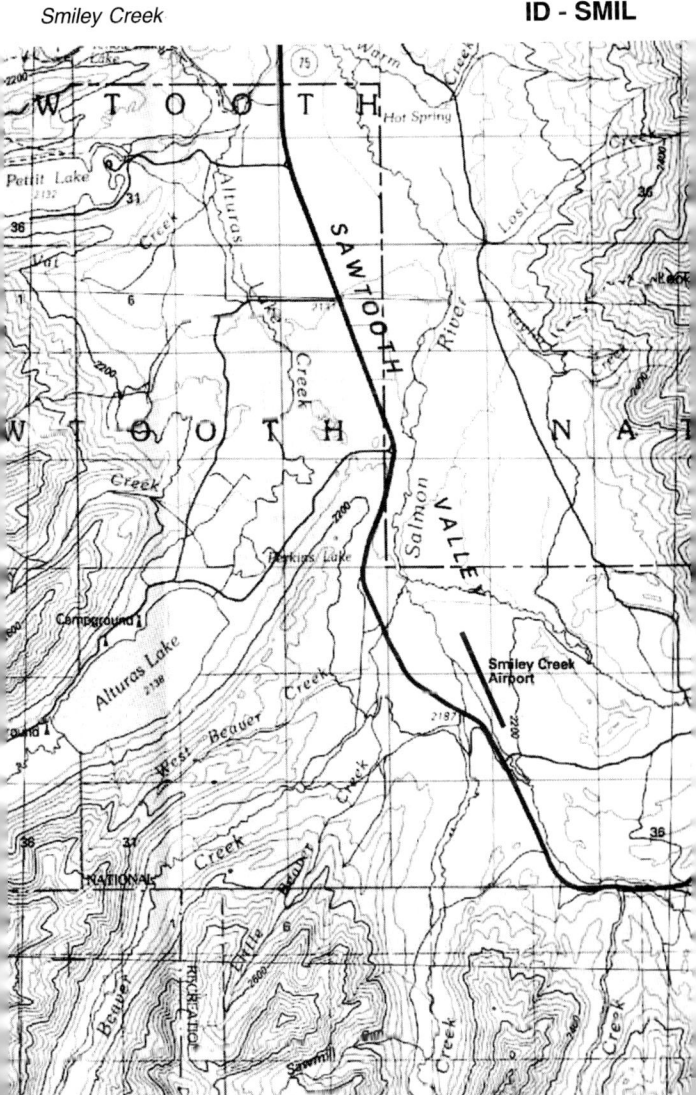

choose your own direction and destination. The topography is varied, providing a number of interesting options.

From this point, you navigate on your own until you again connect with a marked trail. When I visited, I hiked until I found the perfect campsite. The next day, I retraced my steps to the trailhead.

Other Backpacking Loops

With the rental car or a bicycle, you can visit nearby mine sites. Vienna is an abandoned mining town on Smiley Creek. Sawtooth City is on the west bank of Beaver Creek. The old structures are mostly gone, but the scenery is still there.

Unlike the old mine buildings, the Pole Creek Ranger Station still stands. Built in 1909, it is one of the oldest USFS structures in the area. All three of the above sites are within a half-dozen or so miles of the airstrip. Ask someone at the lodge for directions.

Trailheads for a number of other quality backpacking experiences can be reached within twenty miles of Smiley Creek airport. A popular loop is the Toxaway Lake Loop. Trailheads are located at Pettit Lake, just five miles north of the airport. The Toxaway loop is a picturesque one- or two-nighter.

Due to the rugged topography of the Sawtooth Mountains, most backpacking loops require at least two or three days. For further tips, contact the Stanley Ranger Station at 208-774-3000.

- RW

© 2001 by **ALTA** Research

ID - STAN

Stanley

Stanley Municipal Airport

Town	Stanley **ID** 2U7
Coord	N-44-12.51, W-114-56.07
Elev	6403 feet, S end high, 1.3%
Runway	17 - 35, 4300 x 150' turf
Freq	CTAF-122.9
Chart	Salt Lake City sectional

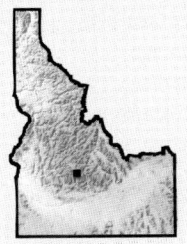

CAUTION: Airport information not for navigational use. Terrain elevation rises to the north; high trees at north end. Soft when wet. Watch density altitude.

STANLEY AIRPORT — STANLEY

AIRPORT LOCATION

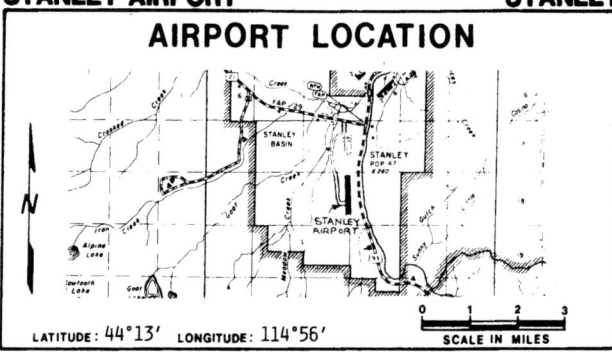

LATITUDE: 44°13' LONGITUDE: 114°56'

AIRPORT LAYOUT

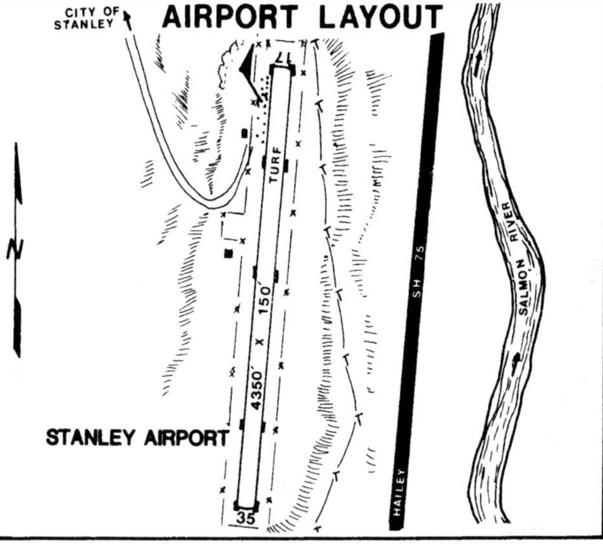

STANLEY AIRPORT

Reprinted from Idaho Department of Transportation 1986 Airport Facilities Directory

Stanley **ID - STAN**

> *Stanley is a gateway to the Sawtooth Mountains, offering backpacking, rafting, fishing, and hunting. $75-350 CofC: 800-878-7950, www.stanleycc.org*

Airport Description

The airport is located northeast of the Sawtooth Wilderness Area. When dry, the Stanley runway has a solid, dirt and turf surface. The north end is lower, with a fence and trees as obstructions. Due to the nearby river, Stanley can become covered by fog or low clouds. However, conditions are usually temporary and can change for better or worse in short periods of time. Smiley Creek airport (to the south) is more often clear, and can be considered a potential alternate.

The airport is closed in winter, except to aircraft with skis.

Services, Transportation, and Lodging

Phone: Idaho Aeronautics, 208-334-8775 (not at airport). Stanley Air, 774-2276, supplies 100 LL fuel. Stanley is less than one mile northwest of the airport. Food, lodging, transportation, sporting goods, and topo maps are available in town.

Tiedown spaces are normally available, but can be filled up during peak seasons. The strip is now owned by Idaho Aeronautics, so tiedown fees are no longer required.

The River Rat Express is the prime source of ground transportation in this area. Owned by Dick, Marilyn, and BZ Waite, the R.R.E. is a family-owned operation that provides busing services for most outfitters in the area. They will taxi people to and from trailheads for $10-50. All family members are pilots and competent weather observers. They claim to monitor 122.9 during the day. Contact the R.R.E. by phone at 800-831-8942 or 208-774-2265. Another alternative is the Mountain Village Lodge, which provides shuttle service, resources permitting.

The airport is close to town, and in-town lodging is within walking distance of the airport. Nearly all motels provide airport pickup if arrangements are made in advance.

Three inns have private **natural hot springs** : Haven Hot Springs, Idaho Rocky Mountain Ranch, and Mountain Village Lodge. The Diamond D Ranch is a full-service ranch that will pickup from the nearby Upper Loon Creek airstrip. Their prices include nearly everything: meals, **horseback riding**, and various other activities. Communication is by backcountry radio, and urgent messages can be sent via McCall Air Taxi at 208-756-4713.

Features
- **B** Full breakfast
- **C** Continental
- **H** Historic
- **J** Jacuzzi
- **P** Pets OK
- **R** Restaurant
- **S** Swim pool
- **T** Transportation

Name	Features	Price	Phone
Danner's Lodge Cabins	HPT	$42-130	208-774-3539
Diamond D Ranch	BHJPRST	$300-360	208-336-9772
Haven Hot Springs (55 mi)	JRS	$68-100	208-259-3344
High Country Inn B&B	BHPRT	$149-171	208-774-3665
Idaho Rocky Mtn. Ranch	BDHS	$200-300	208-774-3544
Jerry's Country Motel	PT	$53-82	208-774-3566
Meadow Creek Motel & Spa	JPT	$75-85	800-811-5745
Mountain Village Lodge	JPRT	$66-120	800-843-5475
Redfish Lake Lodge	HRT	$60-150	208-774-3536
Salmon River Lodge	HT	$60-70	208-774-3422
Sawtooth Rentals & Motel	T	$70-110	800-284-3185
Stanley Outpost	PT	$71-109	208-774-3646
Triangle C Ranch	PT	$61-116	800-303-6258
Valley Creek Motel	T	$66-100	208-774-3606

© 2001 by ALTA Research

Stanley **ID - STAN**

Overview

Odds are that soon after you land, the crisp mountain air will stimulate your appetite. The closest **restaurant** from the airport is Sawtooth Hotel, a good bet for breakfast or lunch. The Mountain Village Restaurant offers three good meals per day. The Casino Club serves tasty pasta and steak for lunch and dinner. Pappa Brunee's and Sawtooth Luce's serve pizza.

A walking tour of the local sites is a good anecdote for those recently acquired calories. The **Stanley Museum** is located just a half mile north of town on Highway 75. It is a local pioneer museum with historic artifacts and photographs of significance to the area. The **Sawtooth Valley Pioneer Park** and **Shaw Homestead Cabin** are south of town, just west of the airport.

A car or bicycle is needed to reach the following sights, all of which are within fifteen miles of town. Two unimproved **hot springs** are located north of Stanley on the river side of Highway 75. Basin Creek Hot Springs is 8.3 miles from town between mileposts 197 and 198, and Sunbeam Hot Springs is 12 miles from town between mileposts 201 and 202.

Further north, the **Boot Hill Cemetery** is located between mileposts 202 and 203. At the same location, you will find more historic sites: **Custer Museum** and **ghost town**, the **Yankee Fork Gold Dredge**.

Stanley is the only location in America where three National Forest Byways converge. These are the Sawtooth, the Ponderosa Pine, and the Salmon River Scenic Byways. This is why Stanley has become a major backcountry center for **rafting**, **floating**, **kayaking**, **backpacking**, **riding**, **hunting**, **fishing**,

© 2001 by ALTA Research

cross-country skiing, and **snowshoeing**. A number of outfitters can help you enjoy backcountry activities:

Blackadar Boating	*raft & kayak rentals*	208-756-3958
Deadwood Outfitter	*pack, fish, hunt*	208-462-3751
Echo: Wilderness Co.	*float trips*	800-652-3246
Mackay Wilderness River Trips	*rafting*	800-635-5336
Middle Fork Rapid Transit	*float trips*	888-433-5628
Mid. Fork Riv. Expeditions	*floats, fish*	800-801-5146
Mystic Saddle Ranch	*pack, hunts*	888-722-5432
Outdoor Adventures	*rafting*	800-323-4234
Pioneer Mtn. Outfitters	*pack, fish, hunt*	208-774-3737
Sawtooth Mtn. Guides	*hike, climb, fish*	208-774-3324
Sawtooth Rentals	*floats, kayaks, bikes*	800-284-3185
Sawtooth Wilderness Outfitters	*pack*	208-462-3416
The River Company	*day raft trips*	800-398-0346
Triangle C Ranch...	*floats, trout fishing*	800-303-6258
Venture Outdoors	*llama treks, mtn. bike*	800-528-LAMA

Although the list of local outfitters seems large for a small town, outfitters quickly become booked for the summer. Advance reservations should be considered a must. The cost of professional outfitting services varies somewhat from no-frills excursions to wine and cheese safaris. When checking rates, be sure to ask what frills are included (Ripple wine or French Champagne, for example).

You can obtain further information and a summary of upcoming events from the Stanley Chamber of Commerce, 800-878-7950. Contact the Stanley Ranger Station at 208-774-3000 for information about permits and hiking in the Sawtooth Wilderness Area.

- RW

© 2001 by **ALTA** Research

Sulfur Creek Ranch

ID - SULF

Sulfur Creek Airport

Town	ref. Cascade **ID** ID74
Coord	N-44-32, W-115-21
Elev	5835 feet, W end high
Runway	8 - 26, 3400 x 50' turf
Freq	CTAF-122.9
Chart	SLC & Great Falls sectionals

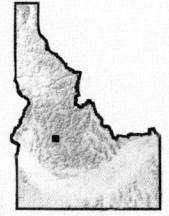

CAUTION. Information not for navigational use. One-way, marginal go-around strip. Land 26; depart 8. Watch for animals on runway.

> *Sulfur Creek Ranch is a wilderness guest ranch in the River of No Return Wilderness. Horseback riding, fishing, and hunting. Fly-ins welcomed for meals. $120-240*
> *Info: 208-377-1188*

Like most Idaho wilderness airstrips, the ranch and runway are in a valley, surrounded by mountains. Plan a safe approach that will get you down on the first try. Go-arounds near touchdown are not possible for some craft. Wind typically changes from west to east after 11:00. Pilots always land on Runway 26 and depart on 8. The runway surface is dirt and gravel.

Airport phone: 208-377-1188 (not at ranch). The ranch has emergency communication by backcountry radio to Arnold Aviation in Cascade. No fuel or phone. Breakfast, lunch, and dinner are served to fly-in visitors or overnight guests.

Sulfur Creek Ranch   **ID - SULF**

There are no roads to the ranch. The only way in or out is by foot, horse, or airplane. Horseback is the mode of choice when exploring outside the ranch. Lodging is provided in log cabins, which have been renovated to include all the amenities.

Places like the Sulfur Creek Ranch are rare in this age of technology, taxes, and regulation. Pilots and their passengers are among the privileged few who can experience deep wilderness, yet enjoy three hot meals and a cozy place to sleep.

Tom Allegiezza, a retired chiropractor, purchased the 160 acre ranch over two decades ago. The facilities remained rough for a number of years. Since then, the cabins have been chinked, paneled, and remodeled. Cabins contain wood stove, hot and cold running water, shower, and toilet. Thanks to the ranch's micro hydro power generator, the rooms have electric lighting, and the beds even have electric blankets.

The ranch's lodge is the center of action. Guests gravitate to the lodge for meals, card games, pool, reading, or a lively conversation. Breakfast is served 7:00 until 11:00. The ranch is

known for a generous breakfast spread, and its reputation attracts hungry pilots who fly in early when the air is good. Meal guests pay $9 for breakfast, $9 for lunch, and $11 for dinner. Dinners always include a main course like steak, prime rib, or salmon. Meals are included in the package for overnight visitors. The daily cost for meals and lodging per person is $60, and less for children. Add guided **horseback riding** to the package, and the price rises to $120.

Thanks in part to the ranch's isolation, nearby rivers and lakes provide **exquisite fishing**. Fishing, particularly fly fishing, has become a prime attraction at the ranch. Right on site, they have a trout pond that produces trophy-size fish.

When **hunting** season arrives, hunting becomes the prime focus at the ranch. You can hunt with a guide for about $3000 per week per person. A spike camp costs $1200 per week. Hunters can expect to bag deer, elk, or bear.

The Sulphur Creek Ranch is just a short flight from Bruce Meadows. See **ID-BRUC** for a fun day-hike to hot springs.

ID - THOM

Thomas Creek

Thomas Creek Airport

Town	ref. Stanley **ID** 2U8
Coord	N-44-43.58, W-115-00.21
Elev	4400 feet
Runway	3 - 21, 2100 x 75' gravel
Freq	CTAF-122.9
Charts	Great Falls sectional

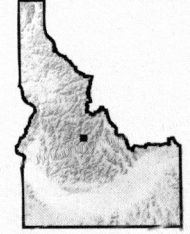

CAUTION: *Airport information not for navigational use. Tricky drafts caused by narrow canyon and tall mountains. Land 21; depart 3. Go-around not possible for some aircraft. Not for inexperienced mountain flyers. Watch density altitude!*

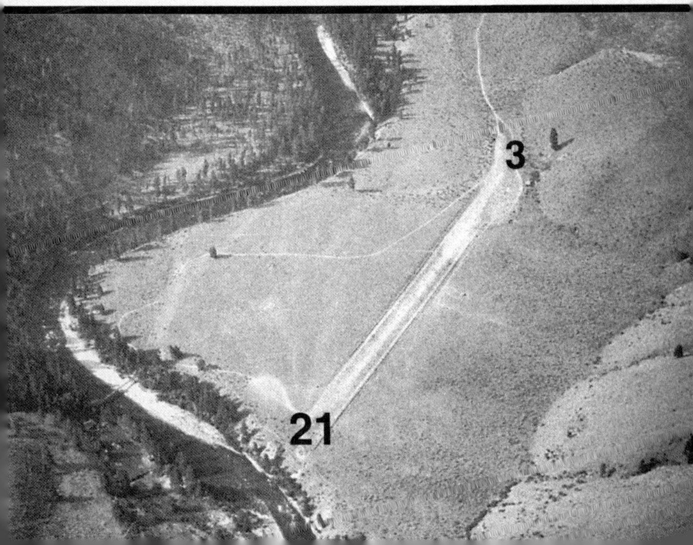

STATE OF IDAHO — THOMAS CREEK

AIRPORT LOCATION

LATITUDE: 44° 44' LONGITUDE: 115° 00'

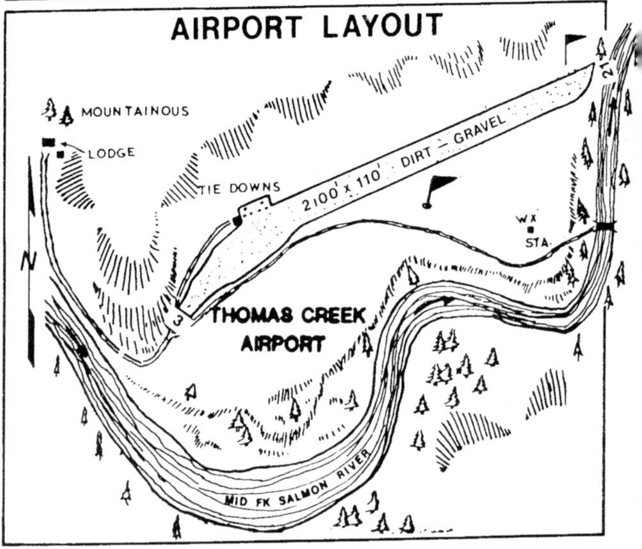

AIRPORT LAYOUT

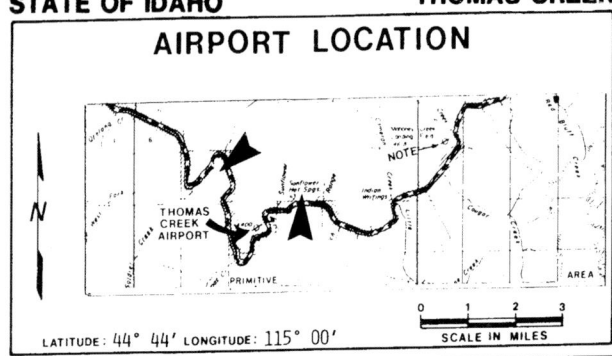

Reprinted from Idaho Department of Transportation 1986 Airport Facilities Directory

Thomas Creek **ID - THOM**

> ***Thomas Creek offers wilderness trails, two natural hot springs, and lodge within walking distance. The strip is surrounded by designated wilderness and inaccessible by land vehicle.***
> ***0$ Challis USFS: 208-879-4101***

Airport Description

The Thomas Creek air strip is located 31 miles north of Stanley within the southern edge of the River of No Return Wilderness. The runway is adjacent and north of a bend in the Middle Fork of the Salmon River. Tall mountains rise on either side of the canyon. Runway surface is dirt, gravel, and small stones.

I need to emphasize that Thomas Creek is appropriate for experienced mountain flyers only. A last-minute go-around is not possible for most aircraft. Fly light.

Land Runway 21. Depart Runway 3, preferably in the morning while the air is still cool. Follow the river northeast until safe to turn.

Services, Transportation, and Lodging

Phone: Idaho Aeronautics, 208-334-8775 (not at airport). No fuel, services, or transportation are available. Camping is permitted, and idyllic unimproved sites can be found on the river and by the nearby hot springs.

The Middle Fork Lodge is a 25-minute walk west from the airstrip. This beautiful, isolated upscale facility has been closed off and on, but has been recently reopened to the public.

© 2001 by ALTA Research

Thomas Creek Northeast Hot Springs

Due to confusion over the years in the naming of the Thomas Creek hot springs, I will refer to the springs by their location relative to the airstrip: "Northeast" and "Northwest."

The Thomas Creek Northeast Hot Springs are unimproved wilderness hot springs, located just 13 minutes northeast of the air strip. They feed a riverside soaking pool. Follow the trail from the northeast end of the airstrip along the river to the springs. Several minutes before reaching the springs, you pass an abandoned cabin on your left.

The spring's source is distributed over a wide area, several hundred yards north of the river. This water is far too hot for bathing, 149 degrees F. The network feeds a man-made pool, one hundred yards north of the river. In my opinion, this pool is too hot for full immersion, but a great cure for tired feet. The most pleasant temperatures are found where the hot water joins the Middle Fork of the Salmon River. Look for a telltale circle of rocks at the river's edge.

You can pitch a tent on soft, level ground less than 100 feet from the riverside pool. With the exception of periodic visits by rafters, your stay will be quiet and private. Nude bathing is the norm.

After finding a perfect spot in the pool, lean back and listen to the river. Let hot water heat your right side, while the cool Salmon River teases your left. Enjoy the majesty of the mountains that reach far above you for the sky. After dark, the waters are like a poem. They made me sleepy, but I did not want to trade sleep for the wonderful sensations.

Thomas Creek Northeast Hot Springs photo, next page . . .

Thomas Creek Northwest Hot Springs

The Thomas Creek Northwest Hot Springs require a 50-minute walk, the latter half of which may be awkward for hikers with large packs. Even if you opt to camp at the other springs, visit the Northwest springs.

Northwest hot springs

Thomas Creek **ID - THOM**

Try to visualize several pools perched on a mountain side with a view of the river. 109-degree water spills from these pools over a ledge and into a natural tub ten feet below. You sit in the rock tub and adjust your body so the hot falls gently pound your back and head. When you've had ten times more pleasure than a person deserves, you dive into the river, not twenty feet away. Oooooo!

Ok, I confess that the springs are not as heavenly as I have led you to believe. I clearly remember sharp rocks poking a rib and butt-muscle as I sat under the water fall. And, yes, the river is a tad cold for my tender hide. Whatever you do, don't tell your friends about this place!

For those who insist on having a look for themselves, here's how to get there: Grab some bug repellant and walk 25 minutes west on the dirt road to Middle Fork Lodge. Cross the river on the lodge's bridge. Do not turn left toward the lodge; the lodge is private and the owners would like to keep it that way. Turn right and follow the west bank of the river north for another 25 minutes. The trail follows steep banks and is occasionally blocked by fallen trees.

Middle Fork Lodge

The Middle Fork Lodge never ceases to surprise first-time visitors. Imagine stumbling on the perfect log-hewn resort in the middle of a designated wilderness. Since it is surrounded by wilderness, every nail and log was flown in by Twin Otter when the lodge was built.

The lodge has top-quality facilities, its own hydroelectric plant, and the grounds are kept bright emerald green. Hot springs water feeds the spa and pool. The grounds have been meticulously maintained. It has been recently acquired by new owners and reopened to the public.

During summer, guests can **fish** at the river's edge, go **horseback riding**, **shoot clay pigeons**, **swim** in the naturally heated pool, and play **tennis**, **horseshoes**, **croquet**, **ping pong**, or **billiards**. Cold weather brings **cross-country skiing**, **snowshoeing**, **hot tubbing**, or simply relaxing by the fireplace in the lodge. The Middle Fork Lodge can connect guests with a licensed outfitter for **hunting trips** further into the backcountry.

For more information, call the lodge's Boise phone at 877-468-635, or see *www.middleforklodge.com* .

- *RW*

Middle Fork Lodge

ID - WARM

Warm Springs

Warm Springs Airport

Town	Lowman **ID** 0U1 (zero-U-one)
Coord	N-44-08.53, W-115-18.86
Elev	4831 feet, N end high
Runway	2 - 20, 2850 x 135' turf
Freq	CTAF-122.9
Charts	Salt Lake City sectional

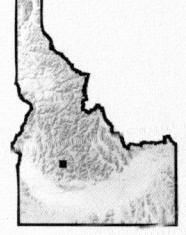

> ***CAUTION:*** *Information not for navigational use. Mountain airstrip, which requires non-standard pattern. Possible high grass and critter holes.*

LOWMAN ØU1 WARM SPRINGS

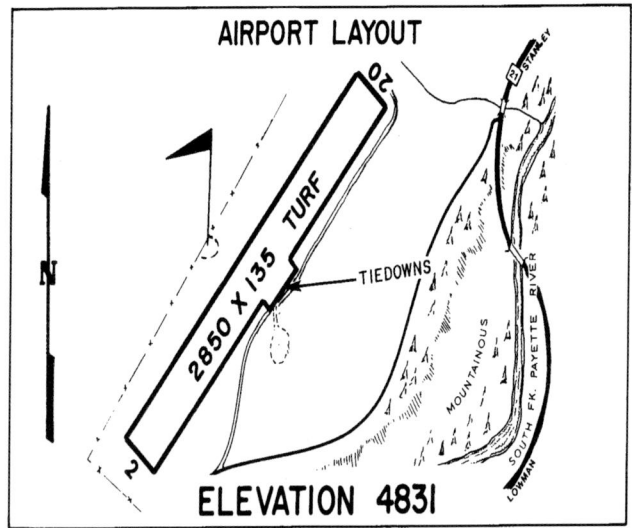

LAT 44° 08.5' N
LONG 115° 18.8' W

VOR FREQ RAD NM
BOI 113.3 029° 53 NM

FSS BOISE (BOI) 1-800-632-6582

Reprinted from Idaho Department of Transportation 1990 Airport Facilities Directory

Warm Springs **ID - WARM**

> ***Fly-in to the Warm Springs pilot's campground and enjoy hot springs, nearby river, and trailheads to the mountains. $0***
> ***Idaho Aeronautics: 208-334-8775***

Airport Description

As with most mountain airstrips, you navigate between the mountain peaks when approaching the strip. This means planning an approach and bleeding altitude at a suitable place in the valley. I chose to approach from the southwest, inspect the runway, circle back and land on Runway 2. The windsock is in the middle, at the western edge of the runway.

After you are setup for the approach, Warm Springs is not a particularly hazardous strip. The runway is wide and slopes up from the south end, favoring landings on Runway 2 and departures on 20. Take care not to delay touchdown, as the far north end flattens out and is less effective in helping you stop.

Taxi near the center of the runway to avoid varmint holes and tall grass. The tiedown area is near the center of the runway, on the east side and abeam the windsock. Several good Idaho Aeronautics tiedowns are provided.

Airport Services

Phone: Idaho Aeronautics, 208-334-8775 (not at airport). There are no aircraft-related services at the airstrip. Located in the woods adjacent to the east edge of the runway is a pleasant, pilot campground with water and pit toilet. The scenic strip and no-charge campground are supported by Idaho Aeronautics and the Ponderosa Flying Club.

® 2001 by ALIA Research

Warm Springs Campground and Hot Springs

The **pilot's campground** contains three widely-separated campsites with picnic tables and fire pits. Unless there is a fly-in, this space is more than adequate. However, you may need to share the space with a car-camper, who either likes pilots or wishes to avoid paying camping fees at the nearby campground – most likely, the latter. I had no trouble scrounging enough wood for a classic white man's camp fire.

The pilot's campground is shaded by a canopy of tall Ponderosa pines. The needles provide a soft cushion for sleeping. If the official campsites are all occupied, you will have no trouble finding a comfortable place to pitch the tent.

Camp camaraderie, mountain country beauty, and good hiking are reason enough to visit Warm Springs. But in my view, the nearby **Bonneville Hot Springs** add a crowning touch to this site. After setting the tent, why not head for the springs!

It's only a 23 minute walk, and I suggest bringing a towel, swimsuit, and drinking water. Begin by walking north on the runway. At the north end center of the runway, keep heading north as you cross a small dirt road at the runway's end. Bear slightly left, and within 50 feet, you connect with a trail that heads north toward the hot springs. Follow this trail down to the road, which intersects 10

Warm Springs

ID - WARM

minutes from the tiedown area. At 14 minutes, turn left and follow the road to the car-camp campground. A *Hot Springs* sign points upstream. You should be leaving the parking area at 17 minutes. Follow a path upstream to the springs.

The first clue is an open area with a small wooden shack and steam rising from various areas around the mountain side. Soon, you will spot the pools down by the river. The water comes from the rocks at temperatures as high as 180 degrees Fahrenheit, so don't even think about testing with a toe. If you have no bathing suit, consider using the privacy of the "rustic bathhouse," which contains a tub of 103-degree flowing water.

The pools by the river have been carefully engineered by visitors, like you and me, who have nothing better to do. The pools offer a variety of temperatures and sensations. Particularly stimulating is the stratification effect: 105-degree water

warming the upper surfaces of your body, while 65-degree water chills the butt. Too intense? No problem, simply move some rocks or step into a different pool. After making the necessary adjustments, I leaned back, placed my head on a rock, and listened to the roar of the white water. I wonder why the river is in such a hurry to get to the sea …

When I visited in June, the car-camp campground was nearly filled. Yet, the springs were totally empty of people. At the hint of dusk, a photographer arrived with tripod and camera to catch the play of steam and late afternoon shadows. He captured the moment on film. I will carry the memory for life.

Red Mountain Hike

Hikers often park at the north end of the runway. This is the trailhead for an interesting, strenuous, high-elevation trail that

Warm Springs **ID - WARM**

climbs to **Red Mountain Lakes**. The hike includes the **Link Trail**, which may well provide not only the most spectacular views of any hike in Boise Forest, but also the best vantage point for seeing the entire length of the Sawtooth range. The destination is 15 miles from the trailhead, so plan at least one overnight stay in the high mountains.

Pack as light as possible, but carry enough water for the grueling 2,920 foot climb at the beginning of the trip. Follow the Link Trail northwest from the trailhead. The next four miles are a continuous climb to Eightmile Mountain, near its 7871-foot summit. On the way up, you'll ask yourself questions like, "Did I really need to bring my smoking jacket and slippers?"

At the top, hike north on a prime stretch of trail with magnificent panoramas, including a close-up look at Grandjean Canyon. After eight miles from the start, the trail joins Castro Creek for a couple more miles. Near Castro Creek, you can cross the river on some logs to a large camping area in a clearing, the best on the trail. This is the only site with dependable year round water.

As the trail continues, it bends four miles northwest, climbing to a ridge, and eventually peaking at the trail's high point of 8120 feet. Now it turns west and southwest. Within a half mile of the turn, you cross the upper end of Eightmile Creek. This signals the beginning of lake country. A number of small lakes are within one mile of the main Link Trail. Another mile, and a light trail branches north toward the lakes. If you haven't had enough climbing, there's more. The trail climbs above the lakes, eventually reaching the Red Mountain Lookout at 8722 feet.

The hike will be safer and more enjoyable if you have the following USGS topo maps: *Cache Creek*, *Miller Mountain*, *Bull Trout Point*, and *Eightmile Mountain*. To obtain U.S. Forest Service maps and information, call the Lowman Ranger District at 208-259-3361.

© 2001 by ALTA Research

8   *Idaho*

Montana

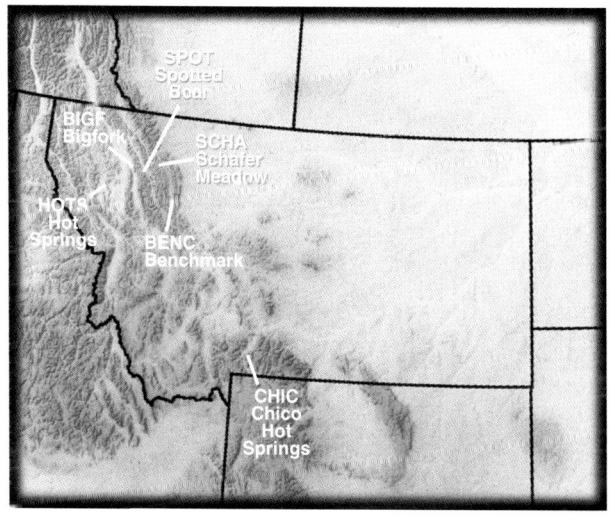

Montana Aeronautics

406-444-2506

Montana

© 2001 by ALTA Research

Benchmark

MT - BENC

Benchmark Airport

Town	Benchmark **ID** 3U7
Coord	N-47-28.88, W-112-52.19
Elev	5434 feet, NW end 75' low
Runway	12 - 30, 6000 x 100' asphalt
Freq	CTAF-122.9
Charts	Great Falls sectional, L9, Lo5

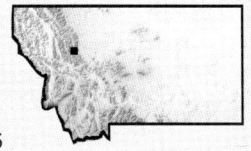

CAUTION: Airport information not for navigational use. Subject to severe crosswinds and turbulence. Mountains nearby. Watch density altitude.

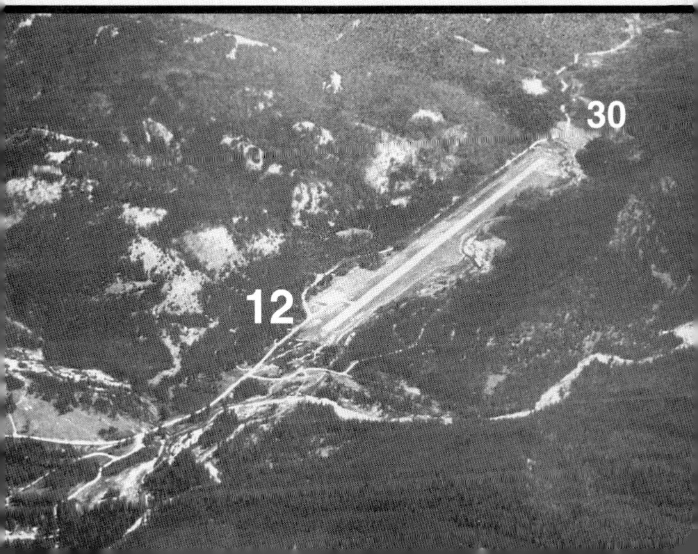

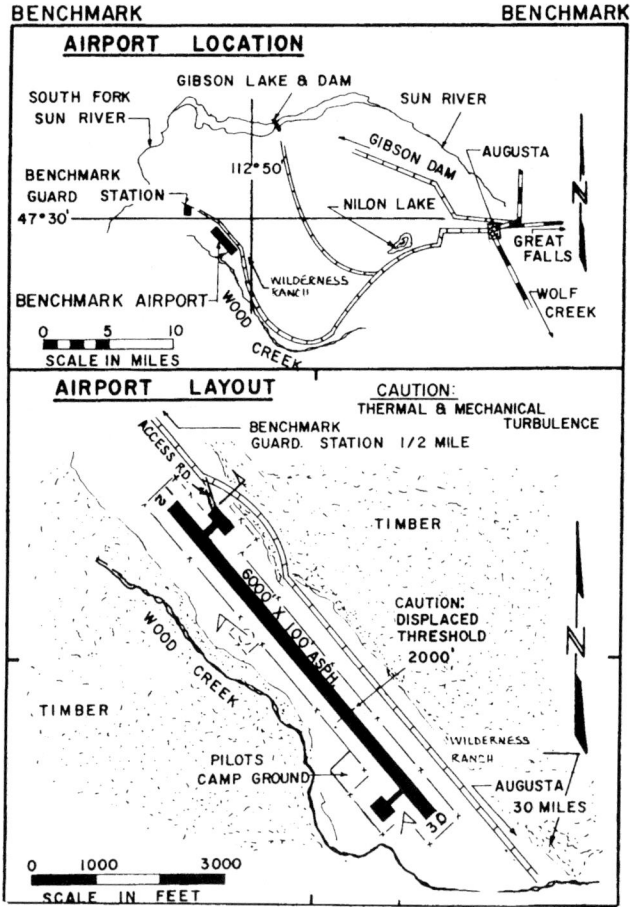

Reprinted from Montana Aeronautics Division 1990 Montana Airport Directory

Benchmark **MT - BENC**

> *Benchmark has an excellent runway, camping area for pilots, and easy access to nearby streams and mountains – a good starting point for multi-day backpacking trips. Wilderness ranch and lodge nearby. $0-270 Ranch: 406-562-3336*

Airport Description

Benchmark is located in a deep valley. Its long paved runway is easy to spot. Windsocks are located at middle and ends of the field. The southeast end is 75 feet higher than the northwest end. The slope is concentrated at the southeast end.

Tiedowns with paved aprons are located at either end. Vigilantly tether your bird, because strong winds can arise without warning. Five tiedowns are available on the asphalt, with extra space on the grass. The pilot's campground is located at the southeast end near the tiedown area.

Services and Lodging

Phone: Ranger Station, 406-466-5341 (not at field). No fuel, services, power, or phone are available at the strip. Cell phones can work, but coverage is spotty.

The Van de Riet **pilot's camp ground** was built by Montana pilots for pilots and passengers. The park features picnic tables, fire grates, water, shade, and outhouse. Wood Creek is located several hundred feet to the west. Area vegetation is a pleasing mixture of sage, evergreens, and aspens.

The **Benchmark Wilderness Ranch** is located one-half mile east of the southeast tiedown. (The sign at the airport is somewhat misleading.)

© 2001 by ALTA Research

Overview

The airstrip is located several miles southeast of the Bob Marshall Wilderness, less than ten miles from the Continental Divide. Trails radiate from the airport area to destinations of geologic intrigue.

Trails to the northwest take **backpackers** into a wilderness area that is home of the **Chinese Wall**. The Chinese Wall is a spectacular geological phenomenon that can be experienced by a several-day backpacking loop. Or, take 20 extra minutes to fly yourself 15 miles northwest for an aerial view of the Wall.

Hikers may obtain the Bob Marshall map for $6 from Flathead Forest Service at 406-758-5200; and area details from the Agusta Ranger Station, 406-562-3247.

Prefer to see the sites by horseback? The Benchmark Wilderness Ranch offers a number of interesting options.

Fly fishermen arrive for a pack trip at the Benchmark Wilderness Ranch

Benchmark   **MT - BENC**

Benchmark Wilderness Ranch

The Benchmark Wilderness Ranch has **guest cabins** and provides **horseback riding**, and **pack trips**. For the sportsman, they specialize in backcountry **big game hunts**, **float trips**, **fly fishing**, **wilderness fishing**, **upland bird hunts**, and **wilderness fishing trips**. Their "**covered wagon family excursion**" would be a premier surprise event for any family.

Although the ranch enjoys catering to pilots and group fly-ins, they cannot guarantee adequate provisions for a large impromptu fly-in breakfast. It's best to make arrangements in advance. There are no phones anywhere near the ranch, so allow them adequate time for communication and the egg run.

Cabins cost $73, a complete package costs $125 per person, and guided trips cost $200 per person per day. Call Darwin or Beverly Heckman at 406-562-3336 for details.

- RW

Guest cabin at the Benchmark Wilderness Ranch

Bigfork and Ferndale

MT - BIGF

Ferndale Airport

Town	Big Fork	**ID** 53U
Coord	N48-04.50 W114-00.06	
Elev	3060 feet	
Runway	15 - 33, 3500 x 95' turf	
Freq	CTAF-122.9	
Chart	Great Falls sectional	

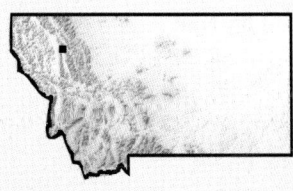

CAUTION: Airport information not for navigational use. Trees and power lines. Field is unusable 4 - 6 weeks in spring, just after thaw.

> *Bigfork is a delightful lakeside town, a vortex for artists and art galleries. Entertainment on weekends includes music, theater, and art gallery openings. $80-250 CofC: 406-837-5888, www.bigfork.org*

Airport Description

As I circled the field for the first time, I pondered "Not much down there. Do I really want to land? Suppose I could pitch a tent and spend a quiet (boring) Friday evening. That bright green turf sure looks inviting. You know, somehow it feels right. Let's go for it!" Reduce power, carb heat, flaps ...

Nestled among tall trees and winding water ways, Ferndale's bright green strip is an inviting target for any pilot with a trace of nostalgia in his or her soul. While circling to land, look for the windsock at midfield on the west side of the strip.

If the area has experienced a recent spring thaw, verify usability in advance by calling 406-837-4113. The field has no lights at night. You can tiedown at Poorman Aviation or the public tiedown area near the eastern corner of the airstrip.

Services, Lodging, and Transportation

Airport phone: 406-837-4113, 257-5994. No fuel and no public phone. Poorman Aviation provides repairs and the kind of friendly support that out-of-town visitors cherish. When I first visited, Poorman's Bob Colby apologized for not having a vehicle I could borrow at the moment. He offered a bike, then

© 2001 by ALTA Research

Bigfork and Ferndale   **MT - BIGF**

paused and excused himself to check on something at his house. When he returned, he handed me his wife's keys, adding "My wife said you can use her car."

At Bob's invitation, I pitched my tent under a wing at Poorman. Facilities include his so called "rude" outhouse, barbecue, and picnic table. The soft, lush turf makes for a perfect mattress.

Recently, Bob Colby and friends have setup a bicycle storage shed containing four courtesy bicycles near the tiedown area. To open the shed lock, you key the airplane's transmitter on 121.5. Clever, these natives.

The airport is five miles from Bigfork. Other options for transportation are: arrange for pickup by your inn keeper, have your resort obtain a car from a Kalispell rental agency, or land at Kalispell and obtain a car at Kalispell. Once in Bigfork, you can

Poorman Aviation at Ferndale airstrip

ride a free trolley to the Eagle Bend Golf Course. Or, rent a mountain bike from Mountain Mike's Rental Bikes, 406-837-2453.

Upscale golfers should consider staying at the beautiful Eagle Bend Golf Course. Town homes cost $650-1000 per three days, pickup from the airport included. The town's trolley travels between the golf course and town at regular intervals. In summer, advance reservations are prudent for this and other Bigfork lodging.

Features
- **B** Full breakfast
- **C** Continental
- **H** Historic
- **J** Jacuzzi
- **P** Pets OK
- **R** Restaurant
- **S** Swim pool
- **T** Transportation

Bayview Resort	**HJPST**	$40-156	800-775-3536
Beardance Inn B&B	**BHS**	$70-150	406-837-4551
Candlewycke Inn B&B	**BJT**	$90-170	888-617-8805
Eagle Bend Golf Course	**T**	$217-333	800-239-9933
Mountain Lake Lodge	**JPR**	$90-225	877-823-4923
Timbers Motel	**J**	$36-42	800-821-4546

Bigfork Inn: fine food and ballroom dancing on weekends

Bigfork and Ferndale **MT - BIGF**

The Candlewycke Inn is just a quarter mile from the airport tiedown area, and three miles from town. The Bayview Resort is located four miles south of Bigfork on Flathead Lake. The resort has a marina, store, and seven cabins with full kitchens. The Timbers Motel is located on Highway 35, just a quarter mile walk from the center of action in Bigfork. Mountain Lake Lodge, five miles south of Bigfork, has two on-site restaurants.

Bigfork

When I rolled into Bigfork, I remember feeling a grin spread across my face. What a surprise! As I rounded the corner, a panorama of shops, colors, tall trees, boats, and water was waiting for me. Though small in size, this lake-side hamlet is rich in **fine restaurants** and **art galleries**. **Specialty shops** and **antique shops** are nestled in between.

This charming corner of Montana attracts a number of accomplished artists whose works are on display in town. Indeed, I met several artists who were performing their magic in the back corners of Bigfork galleries. I arrived on an evening of two art openings. One was an Edible Art Contest at the Bridge Street Gallery and Wine Cafe. I discovered a room full of sculpture that was beautiful, varied, and sometimes very humorous – all made of food! Entry was free. Everyone was invited to cast a ballot for judging the art. Afterwards, we devoured everything.

As hallmarked by the annual Edible Art ritual, fine food is clearly a priority in Bigfork. You will find a number of good places to eat on Electric Avenue, the main street through town. At the north of town, I recommend the Bigfork Inn. After dinner on weekends, you can **ballroom dance** to their **live dance band**. If you like the old big band sound, the total experience at the Inn is like a taste of heaven.

© 2001 by ALTA Research

The Coyote River House is noted for **fine food**, possibly the primo spot in the valley. You can decide! Other notable restaurants and bars include Showtyme (fine food), Waterworks (dining and adjacent bar with music), Garden Bar (good burgers, outdoor drinking, dining, and sometimes music), and the Eagle Bend Restaurant at the golf course.

All flavors of music are available on weekends. **Country and western music** is the specialty of the Tall Pines, northwest of town near the IGA market. **Rock** is played at one or more bars in the Electric Avenue area. Bigfork's **theater** is active through summer months, nightly except Monday. Call the Chamber of Commerce at 406-837-5888 for information about plays and other special events.

Golfers must visit the beautiful Eagle Bend Golf Course. Golf Digest rated it second best resort course in the nation in 1989. Green fees for 27 holes are $45. The facilities include a good restaurant, very nice athletic club, swimming pool, yacht harbor, and condos for rental. Call 406-837-7300 for details.

Flathead Lake is a prime asset of the area, and a number of local outfits make it their business to help you experience it. You can **rent a boat** from the Marina Cay Resort, 800-433-6516. Or, Far West Excursions will take you on a **motor boat trip** around the lake. Questa Sailing Charters invites you to **sail the lake** in their classic Q-class sloop, 406-837-5569.

Sportsmen can enjoy a **fishing excursion** on the lake from Bagley Guide Service, 406-837-3618; or from Able Fishing Charters, 800-231-5214. Both claim "guaranteed fish."

If that's not enough to keep you busy for several days, the Chamber of Commerce can provide additional hints.

- RW

© 2001 by ALTA Research

MT - CHIC

Chico Hot Springs

Chico Hot Springs (road)

Town	Pray	**ID**	none
Coord	N-45-22, W-110-42 est.		
Elev	5280 ft, S end high, 2%		
Runway	14-32, 6000x30' **CLOSED**		
Freq	CTAF-122.8		
Chart	Great Falls (not charted)		

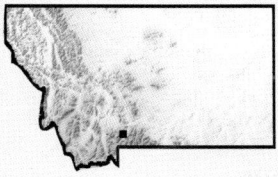

> ***CAUTION:*** *Airport information not for navigational use. The runway was an automobile highway, now **closed to aircraft**. Land at Flying Y, Livingston, or Bozeman.*

> *Chico Hot Springs is a large, historic hotel complex set at the foot of mountains, not far from Yellowstone Park. Known for hot springs facilities and gourmet-quality dining.*
> *$85-250   Lodge: 800-468-9232, 406-333-4933*

Airport Description

Prior to this update, Chico's airport was a county highway that doubled as a runway. The portion of the highway that was approved for landing was just west of Chico Hot Springs, and about three miles east of Pray (a small town of 25, marked on the Great Falls Sectional). The highway runway is now closed to aircraft. Land at Bozeman, Livingston, or the nearby Flying Y airstrip, 406-333-4788.

The landing area was delineated by windsocks at both ends. Daylight operations, only. Due to the crown on the road and depressed drainage areas on either side, landing in winter was highly discouraged. Tail-dragger pilots were advised to be extra sharp any time of year.

To land on the road, pilots were to call Chico in advance at 406-333-4933 to alert them of the arrival. Upon arrival, the pilot would circle the runway and contact Chico Hot Springs on 122.8. If no contract, the pilot would try to get someone else to call them on a land line. After the road was blocked by Chico Hot Springs staff, the pilot would land.

After touching down, airplanes would simply taxi straight ahead and park on the grass in front of the motel. Ah, the good old days!

© 2001 by ALTA Research

Chico Hot Springs **MT - CHIC**

Services, Transportation, and Lodging

A rental car is not required if you plan to stick close to the lodge. For a fee, Chico provides shuttle service to and from the Flying Y. Limo service is also available from Livingston and Bozeman. Once on site, visitors can rent **mountain bikes**.

In Livingston, rental cars are available from Livingston Ford, 406-222-7200; or Livingston Auto, 222-8600. Or, fly 25 miles further west to Bozeman and choose from a number of other rental agencies.

Chico Hot Springs offers a wide spectrum of lodging with prices to match. Hotel rooms cost $48-114, depending upon size and bath facilities. The hotel rooms are typically small and are furnished with old furniture and brass beds, about what you

Chico Hot Springs lodge with road/runway in the background

would expect to find above an old west saloon. The new Lower Lodge near the tiedown area has rooms for $80-100. Six other options range from $130 for various facilities to $327 for the five-bedroom Mountain View Log House that can sleep up to twenty.

Chico's primary facilities are rustic, yet clean and professionally managed. Owner Mike Art comments "You'll find very little plastic here." However, amenities such as hot tubs, satellite TV, telephone, and laundry facilities are available at the house rentals mentioned above. Call 800-468-9232 for details.

Chico Hot Springs

Tucked snugly at the knees of Emigrant Peak and Mt. Cowan, Chico's timeworn buildings sit where the flats of Paradise Valley meet Montana's thickly forested mountains to the east. From deep beneath the surface bubbles a non-sulfur mix of hot mineral

The old tiedown location was convenient

Chico Hot Springs **MT - CHIC**

waters, which some believe have healing properties. The story of the white man's use of the springs dates back to 1865 when prospectors used to bathe and wash their duds in nature's laundromat.

The hotel was founded just prior to 1900. As the years passed, it grew to a full-featured four-season, 62-room resort. The resort's future looked dim as the old buildings lost their sparkle through the 60s, but the hotel was rescued by Mike Art and his family in 1973. Mike tossed his career as owner of 36 men's clothing stores in New York, and moved his family to a new life in Paradise Valley.

Years of restoration followed. Perhaps Mike's single greatest accomplishment is the hotel's restaurant. Montana residents literally travel hundreds of miles to select from a wide variety of culinary treats. The decor is western with a touch of New York

The Chico lobby, a place to make new friends or simply kill time

perfection. The service is less personable than the western norm, but the menu is somewhat better. The wine list contains a number of good picks. I selected buffalo as my entree. The cliche "it melted in my mouth" describes the experience exactly.

The dessert tray could have rolled out of one of New York's best restaurants. Let me tell you about the house specialty: An orange is scooped out. The pulp is mixed with sour cream, chocolate, vanilla and three liquors. Meringue is slathered on top; and when the concoction is placed before you, it is ceremoniously doused with brandy and set on fire.

In summary, the restaurant is big-city quality at a correspondingly big-city price – expect to spend $25-50 each. Although children's items are not listed on the menu, they are available upon request. Reservations are highly recommended. Other options include the Poolside Grille or outdoor BBQ.

The **hot springs** includes a large outside **swimming pool** with views of the mountains. Swim suits are required. The large pool is 94 degrees and the smaller is 104. **Private hot tubs** and **massage therapy** are available for a fee. The therapy room provides nearly a half dozen different treatments for $35 per half-hour, $50 per hour, or $70 for 1.5 hours.

Horseback rides are available year round. **Steak rides** require 72 hour notice. Guests may fish at the lodge's **trout lake**, a 20 minute walk from the hotel. Chico can arrange a wide range of outdoor activities: **fishing trips**, **drift-boat rides**, and **scenic excursions**. Fishing instructors and rental gear are available for the novice.

During winter, the focus changes to **cross-country skiing**. **Absaroka Dogsled Treks** in Pray provides a unique adventure: 2-hour dogsled rides for $70 per person, half-day for $130, and full-day rides with gourmet box lunch for $170.

© 2001 by ALTA Research

After dark on weekends, watch the caravan of local 4x4 pickups roll into the saloon's parking lot. Saturday night gets wild and crazy at the **Chico Saloon**: Live poker, pool, poker machines, video, and **live music on weekends**. It's quite a show, and beers are sensibly priced. Check your guns at the bar.

Walk to Old Chico

Old Chico is a semi-ghost town about two miles further east from the resort. I chose to walk to Old Chico by moonlight. After a 40-minute walk, I was somewhat disappointed to find the town more occupied by real people than ghosts. Still, the buildings were old and I liked the roar of the river and the feel of the night. The road led me and the river up to a canyon, until darkness convinced me to return.

Moon-lit cabin in old Chico (time exposure)

On the way back, two white Siamese cats glowing of moon sprang out in front of me. They continued to zigzag in front of me, occasionally becoming tangled in my feet. As quickly as they came, they disappeared into the weeds. At the next bend they sprang at my feet, clearly loving their play with their quarry. White playful blurs with black tails, full of moonglow, the evening breeze rushing down the canyon . . .   Yes, another perfect evening in Big Sky Montana.

- RW

MT - HOTS

Hot Springs

Hot Springs Airport

Town	Hot Springs
ID	S09 (S-zero-9)
Coord	N47-36.75, W114-36.81
Elev	2763 feet
Runway	6-24, 2580 x 40' oil/gravel
Freq	CTAF-122.9
Charts	Great Falls sectional

CAUTION: Airport information not for navigational use. Powerline and fence. Runway can be rough or muddy.

> *Hot Springs, a sleepy community with several old-time spas, and outdoor pools and mud baths – a funky scene that's gentle on the pocket book. $60-100*
> *Symes Hotel: 406-741-2361*

Airport Description

The airport is easy to locate. It sits on the flat valley floor near a river and two miles east of Hot Springs. Before landing, inspect the runway surface if there is any chance that the soil has been saturated by heavy rains.

Airport Services, Transportation, and Lodging

Airport phones: Montana Aeronautics, 406-444-2506 (not at airport); caretaker, 741-3582. The airport is not officially attended. Cell phone coverage may be spotty (depending upon your carrier), but there is now a public phone on site. The caretaker at the adjacent highway maintenance yard may be able to drive you into town. The most reliable procedure is to arrange transportation before departing for Hot Springs.

You can pitch a tent at the airport, but there are no rest rooms and the environment is not attractive. Camping is allowed for a fee at the Camas Hot Springs in town, and at the Wild Horse Hot Springs. The following inns pickup from the airport if arrangements are made in advance. Once in town, everything is within walking distance.

© 2001 by ALTA Research

Hot Springs **MT - HOTS**

The Symes Hotel, 406-741-2361, is my favorite. It's an old-fashioned place, typically with no TVs or phones in the rooms. The facilities have been recently renovated, yet maintain a historic atmosphere. As you walk into the lobby, you feel like you are stepping back in time.

Rooms are available with and without baths, $57-85. Cabins can be rented by the week for $47 per day. Their private Jacuzzi suite costs $95. Baths in rooms are piped with hot mineral water. And now, guests can enjoy a soak their new 104-degree outdoor pool.

A masseuse is available daily, providing body work for $60 per hour. Breakfast is served every morning, and hot soup is always on the stove. On weekends, the restaurant goes full service, providing a variety of musical entertainment in the evenings. Mountain bikes are available for rental by guests.

Symes Hotel, remodeled since this photo

The Camas Hot Springs Spa, 406-741-2283, is located a few blocks closer to the defunct outdoor public pools and mud baths. The facility is funky, yet tolerable for visitors who can adjust their expectations. For $26-47, you get a room with cable TV and kitchenette. A resident masseuse works her magic for $35 per hour. They pickup from the airport on prior request.

Wild Horse Hot Springs, 406-741-3777, is five miles northeast of town. You can bring your own food, or the owners will drive you to a restaurant in town. They have clean private bathing rooms, each with toilet, shower, and steam room. The spa is open from 9:00 AM to 9:00 PM, and prices are $5 per person per hour.

Overnight guests have the choice of staying in various rooms or pitching a tent on green grass in the RV area. A room with one

One of five pools at the free, public hot springs area   (now closed)

Hot Springs **MT - HOTS**

queen costs $55, and two queens costs $65. Make arrangements in advance, and buzz the facility for pickup.

Public Hot Springs

The free public hot springs, once an attraction at the north end of town, are now but a historical artifact. To get there from the Exxon station, follow Spring Street about four blocks north. At the Y, make a hard right into a gravel parking lot. A defunct concrete pool is at your right. The gravel road takes you to the remains of the pools.

There were once five pools of varying character and temperature. One mud bath was open and the other was covered with a roof. The mud soakers consisted of stall-like units that allowed

Soaking in hot mud doesn't have to be a serious matter (closed)

you to bathe in water while sinking your feet into warm mud. Strangely, the mud did not stick to the body.

The town of Hot Springs is built on the Flathead Indian Reservation. The Confederated Salish and Kootenai Tribes, owners of the hot springs property, had recently installed a concrete pool for public use. While they had no mandatory fees, they asked that visitors donate what they could to help pay for insurance.

Whether it was due to the cost of insurance or to a half dozen other reasons that have circulated as rumors, the tribes have shut down the hot springs.

An alternative is to visit the Camas Hot Springs on the hill north of the now-closed public springs. Access to the springs costs $5 per person, and camping is allowed.

The **local restaurants** are typical small-town restaurants; food is basic and palatable, at a reasonable price. As mentioned earlier, the Symes Restaurant serves breakfast every morning, and three meals on weekends. Try the Red Tail Cafe for a good breakfast or nice lunch. The Running Iron is another alternative in town for breakfast and other meals. The Second Home Restaurant, not far from the airport, serves good lunches and dinners.

In the evening, restaurants close early. If the cafes are closed, try the Pioneer Bar's broasted chicken and jo-jos. They also serve homemade pizzas.

Downtown **night life** is limited to low-cost drinks and bar TV. I suggest that you bring a good book or playful companion.

- RW

© 2001 by ALTA Research

Schafer

MT - SCHA

Great Bear & Bob Marshall Wilderness

Schafer USFS Airport

Town	Schafer Ranger Station
ID	8U2
Coord	N48-05.25, W113-15.05
Elev	4855 feet, W end 25' low
Runway	7 - 25, 3200 x 60' turf
Freq	CTAF-122.9
Charts	Great Falls sectional

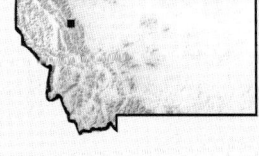

CAUTION: Airport information not for navigational use. Trees and high terrain east end. Tall mountains in all quadrants. Fence, standing water, and livestock. Watch density altitude.

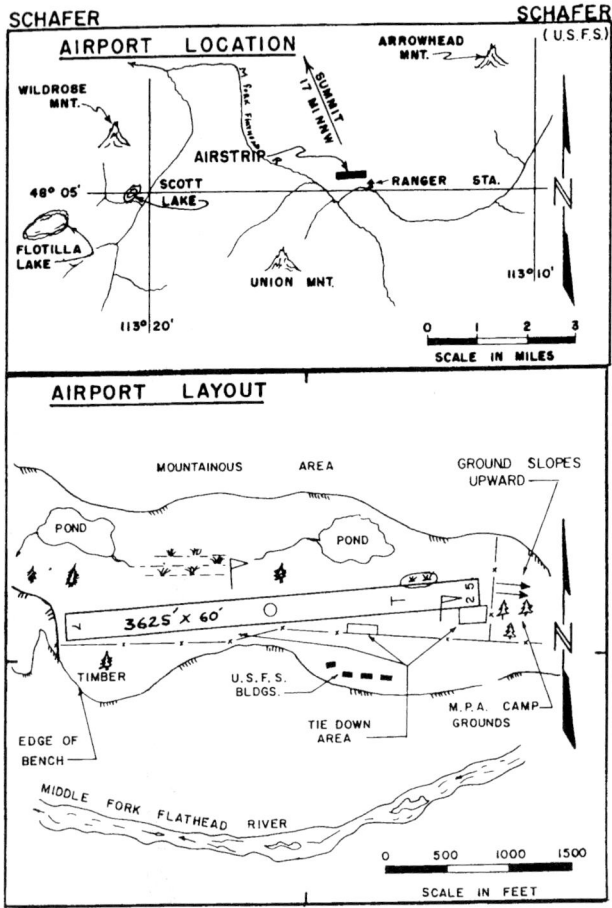

Reprinted from Montana Aeronautics Division 1990 Montana Airport Directory

Schafer (wilderness) **MT - SCHA**

> *Schafer, a mountain airstrip with a great pilot's camping area. Favorite haunt of Montana pilots. Fish, hike, or kick back and enjoy tall tales. $0 USFS: 406-758-5200 or 406-755-5401*

Airport Description

Schafer is located inside the Great Bear Wilderness, 15 miles from the closest road. First, locate the Flathead River's Middle Fork. Once you have the river, the airstrip is easy to see. Two bright orange windsocks are located at either end of the runway.

Due to Schafer's wilderness status, touch and goes are not allowed. If you land, you are expected to stay over night. These rules are necessary to keep traffic below Federal LAC (Limits of Acceptable Change) minimums.

Land and depart on Runway 25. When landing, fly close to the trees on final approach; otherwise, you may need to execute a go-around. Go-arounds are relatively safe in the westerly direction, but not recommended to the east. The turf runway is usually in good condition, but soft field conditions are possible after extended precipitation. To prevent prop damage, **watch for critter-holes when taxiing off the runway**.

Tiedowns are installed at two locations: at the east end near the campground and at midfield near the ranger station gate. Choose a tiedown at the northeast corner if you want evening frost to be melted by the morning sun.

Although a number of airstrips still exist inside wilderness boundaries, Schafer is the only Montana site left for general use. It also serves the Forest Service and is an embarkation point for outfitters and rafters.

© 2001 by ALTA Research

Services, Lodging, and Transportation

Phone: USFS, 406-758-5200 (not at airstrip). Ask for "dispatch." Or, try 406-755-5401. Neither phone, power, nor aviation services are available at the strip.

Although no formal lodging is available at Schafer, a wonderful campground is provided for pilots at the east end of the strip. Facilities include picnic tables, fire grates, log-constructed outhouses, and water. Fire wood is available close by. The sites are shaded by trees, just a short walk from the tiedown area.

A fire circle, benches, and water spigot are located just under final approach for Runway 25. This area provides a sunny community center where pilots sip coffee, spin yarns, and critique landings. The facilities are provided and maintained by Montana Aeronautics and local pilots.

Schafer Area

Without qualification, I love this spot; it promises to be a regular stopover point for me. The campsites are perfect, hiking trails radiate in all directions, and the pilots that visit are a kick in the pants. Become a Schafer regular, and someday we will meet – maybe at the annual work party, the third weekend of July.

Let me share the details of my first visit to Schafer. I often travel to wilderness areas with light (dehydrated) camp food. Hiking in the wilderness followed by a light meal is my answer to a diet. Great theory, but the plan doesn't work at Schafer.

The evening I arrived at Schafer, a band of local pilots was cooking up a banquet of fresh vegetables and thick juicy steaks. Dave said, "Reed, you want a steak? We always bring an extra, 'case someone like you arrives." How could I refuse? With the exception of a buffalo steak enjoyed at Chico Hot Springs several days earlier, the Schafer Special tasted better than any restaurant steak in recent memory.

© 2001 by ALTA Research

Schafer (wilderness)

Evening entertainment included nonstop stories around the campfire. The unadvertised theme for the evening became "true" stories.

Herman set the pace. He began: "Did I ever tell you how I got me a moose with a BB gun? Now this is a true story. As a kid I spent my summers in the mountains by a pond. One summer there was this damn bull moose, he come down there and feed in that pond. You know how they feed: put his head down in the water for a long pull, and bubble air out his nostrils, raise his head up, look around. I watched that bastard all summer long! And it was during the depression when things were pretty rough – lucky to eat anything." Laughter.

"I was trying to figure out how in hell I'm gonna get that moose all summer long. All I had was a damn BB gun! And, watchin' him all summer long, just on time every morning.

"First goddam morning of hunting season I went out there. Every time he stuck his head under the water I'd get a little

A pilot needs a hearty breakfast after an evening of spinning yarns

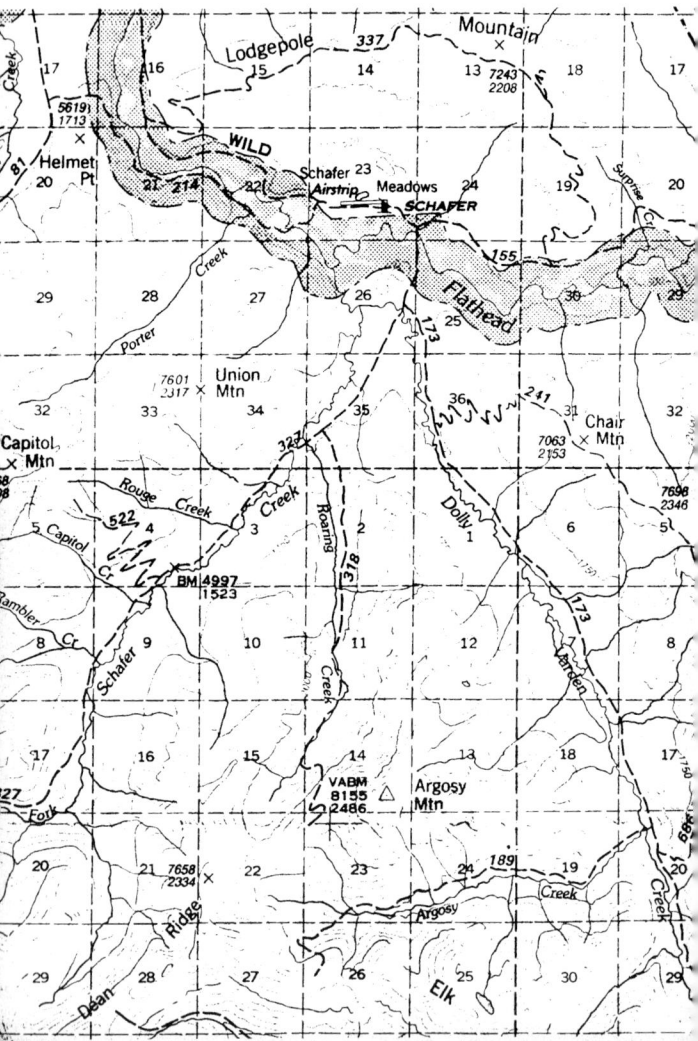

Schafer (wilderness) **MT - SCHA**

closer, cause with a BB gun you have to be pretty close." Much laughter. "Finally I got close enough and he put his head down under the water and I up and shot him right between his legs, and he drowned." Several minutes of laughter.

Chair Mountain Hike

The area offers a number of scenic hikes that can be looped in a day or less. You can purchase maps for the area from the Flathead Forest Service, 406-755-5401 or 406-758-5200 .

The following hike to the summit of Chair Mountain can be walked in an afternoon. Elevation gain is 2000 feet and round-trip distance is less than ten miles.

Head southwest from the ranger station, taking Big River Trail. Shortly transfer to the Schafer Creek Trail and cross the

View of Schafer from Chair Mountain

Flathead River's Middle Fork. There is no bridge, so you must wade across. Connection with the trail on the other side is not obvious. Walk south along the river for four minutes and follow the trail into the woods.

Thirty-five minutes after leaving the trailhead, you intersect the Dolly Varden Trail. Follow it on the left side of the Dolly Varden River. The trail climbs away from the river and at 52 minutes (from start), turn left on the trail marked "Chair Mtn 241." Running water from a spring is located 10 to 20 feet to the right of the new trail. Climb the shaded trail through 13 switchbacks toward the peak. At two hours at elevation 6750, the trail opens for panoramic views. Near the summit, break left from the trail and hike to the 7063-foot summit. Total time up is two and one half hours.

The return trip is much quicker. When crossing the Flathead Middle Fork, look for a one-foot-square yellow sign with a black "X" that marks the crossing point.

"Don't just stand there Herman! We're not leavin til you find a scrap of wire."

MT - SPOT

Spotted Bear

Spotted Bear Airport

Town	Spotted Bear **ID** 8U4
Coord	N47-57.50, W113-33.56
Elev	3670 feet
Runway	14 - 32, 3800 x 78' turf
Freq	CTAF-122.9
Chart	Great Falls sectional

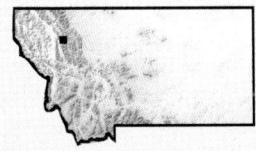

> ***CAUTION:*** *Airport information not for navigational use. Trees. Overruns at either end are very rough. Gopher holes adjacent to runway.*

SPOTTED BEAR

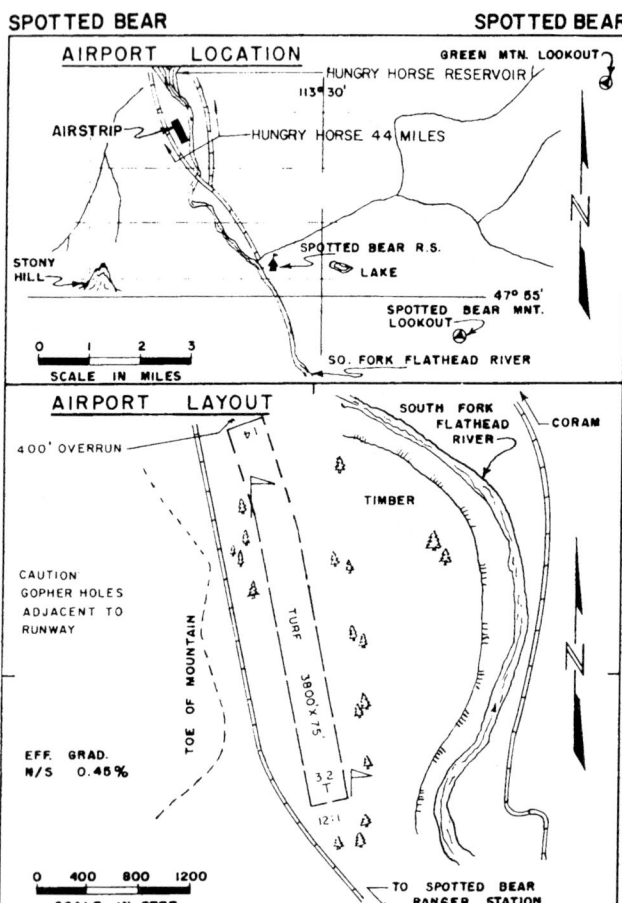

Reprinted from Montana Aeronautics Division 1990 Montana Airport Directory

Spotted Bear **MT - SPOT**

Spotted Bear, a wilderness strip with three lodges. All pickup from the strip. One within walking distance provides ranch-style meals. Unimproved riverside camping near the runway. $0 - $300

Airport Description

The airstrip lies west of a river at the southern tip of the Hungry Horse Reservoir. The north end of the strip has a dog-leg that ends at the river. Mountains rise to the east and west, but most mountain flyers have no difficulty setting up an approach.

Spotted Bear is a USFS non-lighted turf wilderness strip that is not guaranteed to be in good condition at all times of year. It is officially closed November through April, but was in very good shape when I visited in late summer.

Typically, you land south. Taxi to the south end to tiedown and use your own tiedown gear. Conditions permitting, depart north. When departing, follow the strip's dog-leg to the left at the far-north end.

Airport Services and Transportation

Phone: USFS, 406-758-5376 (not at airport). There are no aviation services or fuel at the field.

Local lodges provide transportation to and from the airstrip. The Wilderness Lodge is a 25 minute walk. Hitchhiking works, although traffic in the area is very light. Unofficial riverside campsites can be found five minutes from the southern tiedown area.

© 2001 by ALTA Research

Lodging

You can camp anywhere at the airport, but there are no formal campsite facilities. Better yet, follow the grass road northeast a short distance to the river and pick a scenic campsite at the river's edge.

The **Wilderness Lodge**, a.k.a. **WL Ranch**, is the closest source of meals and lodging. Lodge and nearby log cabins sit peacefully below towering timbers. The innkeepers have a way of complementing the rustic serenity with their easy pace and laid-back conversation.

The Wilderness Lodge is a favorite meal stop for local Montana flyers. If you can get their attention with a little buzz, the folks from the lodge will meet you at the airstrip. Since the lodge is under cover of trees, first-time visitors may have difficulty locating it from the air. Look for the lodge one mile south-southwest of the strip.

The walk to the lodge from the tiedown area takes 25 minutes. Walk west to the gravel road and follow the road south for 15 minutes until intersection and sign "Meadow Creek Trailhead 12 miles". Turn right toward Wilderness Lodge. After another seven minutes you're at the WL Ranch entrance; then three more minutes to the lodge.

You can order a full breakfast for $6, lunch for $6, and dinner for $12. Meals are family style – all you can eat. If you stay, the fare is about $75 per person for

Spotted Bear **MT - SPOT**

three square meals and lodging in one of the seven small cabins. The cabins have bunk beds, bathroom, shower, and optional cat. The six cabins can accommodate a maximum of 29 people. Off-season, prices are reduced for hunters with sleeping bags. Recreation includes **riding, fishing, float trips, hunting trips, pack trips**, and **sight-seeing**, all for additional fees.

The lodge has no phone, but you can make reservations by calling owner Cameron Lee at 800-257-0580 or 406-387-4051. Ask Cameron to explain his package deals.

The **Diamond R Ranch** sits at the edge of a restless river. This ranch is similar in flavor to the WL, but the Diamond R has more units and is a tad more commercialized. Food prices are comparable. Units do not include toilets. Room and board runs about $70 per person.

Wilderness Lodge equestrian quarters

The ranch features full-day **riding** or **fishing** activities. A guide and boat costs $250 per day for two. **Riding** costs $40 per person for 2 hours, or all day with lunch for $100. **Fly-in meals** cost $7 for breakfast, $6 for lunch, and $12 for dinner. Pickup from the airport costs $5 for the ride. Call 406-756-1573 to make reservations and find out about **float trip** plans for pilots. There is no phone at the ranch, but their radio-link may be back in operation by the time you read this.

Just down the road from the Diamond R is the Orvis endorsed **Spotted Bear Guest Ranch**. Of the three, this is the most luxurious and most expensive. The cabins are rustic and relatively large, compared to the guest ranch norm. They are furnished with carpeted floors, refrigerator, and other modern amenities (except no TV or phone).

Owner Kirk Gentry specializes in package deals. For example, three days with meals and **fishing**, **hunting**, or **riding** cost $1275 per person. Five days cost $2125. Another option is their **Wilderness Pack and Float Trip**, which includes a two-day horseback ride into the wilderness, followed by several days of floating. Their hunting trips target black bear, deer, and elk. Call 800-223-4333 or 406-755-7337.

-RW

Oregon

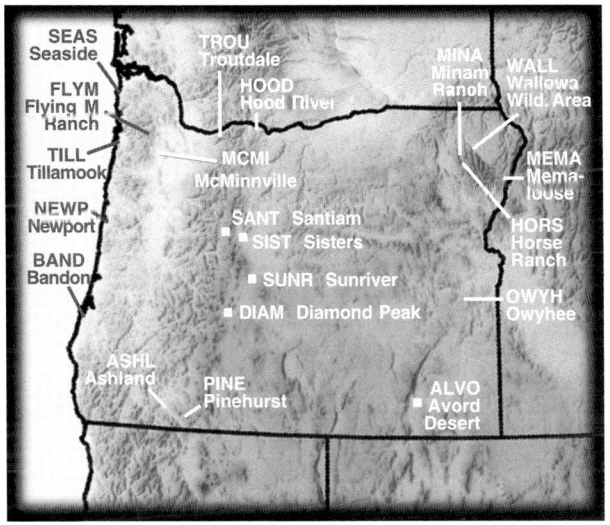

Oregon Aeronautics

503-378-4880

Oregon

© 2001 by **ALTA** Research

OR - ALVO

Alvord Desert

Alford Desert Playa

Town none **ID** none
Coord N-42-31, W-118-29, variable
Elev 4580 estimated
Runway no specific runway
Freq CTAF-122.9
Chart Klamath Falls sectional

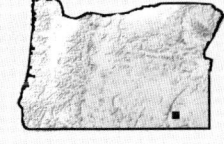

CAUTION: Airport information not for navigational use. The playa is likely to be too soft for landing in spring. Do not land if wet. Inspect surface for firmness.

The Alvord Desert is a playa that can serve as a landing platform in late summer and fall. Hot springs are a short walk from tiedown. $0

Airport

The Alvord Desert contains a salt flat on the eastern edge of the Steens Mountains, which rise abruptly to the west. If the surface were dry and solid enough to handle great weight, it could easily accommodate the landing run-out of a space shuttle. However, even when the surface appears to be dry, it cannot be trusted to carry great weight. The surface is usually safe for small aircraft when dry, typically late summer and fall.

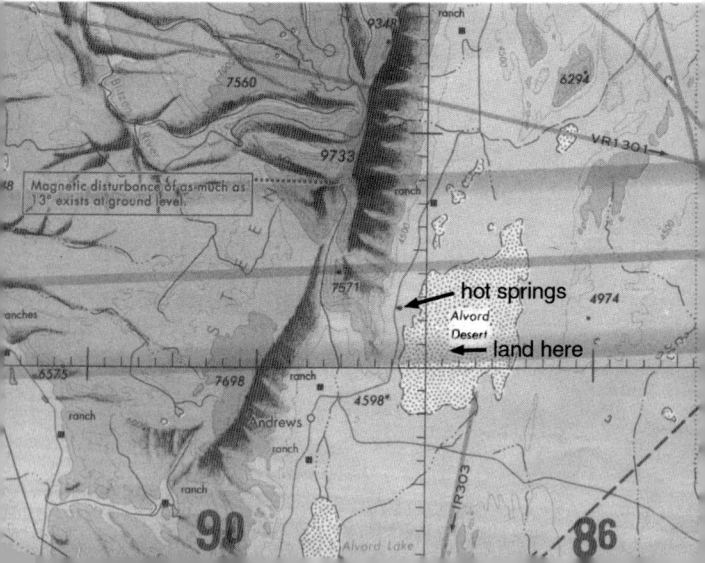

Alvord Desert **OR - ALVO**

Select a place for landing that is most likely to be dry. Reasoning that I was no heavier than the average car, I selected a long stretch with recent car tracks, southeast of the final destination. "Drag" the intended runway before landing. Be ready to apply full power if the surface proves to be wet – the mud on the flat is particularly insidious

A GPS will lead you toward the hot springs coordinates of N-42-32.613, W-118-31.968. If without GPS, taxi west to a spot close to the road. A good access point is located between a lone house to the south and wet flat area to the north. You can safely taxi to within about a mile of the springs.

Some pilots land on the north-south dirt road between the mountains and the springs. If there is any chance that the playa is soft, the road may be the safer option. If you take this alternative, land north of the hot springs because power lines cross the road to the south.

Tied down at edge of playa

Hot Springs

I arrived over the desert flat just before sunset. The sun was already behind the towering Steens Mountain, and daylight was rapidly waning. With mere with minutes to spare, I headed for an area with traces of car tracks.

To verify that car tires had not pierced into soft surface, I dragged the area. The surface looked so good that I just let the bird down during a low drag. Without delay, I taxied for the closest access point to the road, which passes north-south between mountains and flat. Meanwhile, I noticed a couple of latecomers landing in the middle of the flat.

The hike to the springs takes 20 minutes. Walk to the road. Then walk north until the roadside parking areas and the springs, 200 feet east of the road. The springs consist of a deck, open bathing area, walled bathing area, and a roofed, three-wall shelter that could serve as a dry place for two or three people to sleep.

Alvord Desert **OR - ALVO**

The pools are of pleasant temperature, offering but a hint of sulfur smell. Although the pool bottoms are covered with algae, several algae-free seats are provided to keep your tail from turning green. Bathing suits are optional, but be aware that local folks may feel uneasy with nudity.

After a good soak, I walked back to the plane in the dark and snuggled into the bag without bothering to setup the tent. In the desert, sleeping without a tent is almost always a bad idea. However, the desert is miles from the lights of civilization, and I wanted to enjoy the brilliant stars before dozing off.

While admiring the incredible sky, I noticed a bright light heading exactly in my direction from the center of the flat. Oooo, X-Files, I think. With glasses on, I see vehicle headlights, which rolled to a stop ten feet from me and the plane. Turned out to be a gal in a truck who was on a "virtual vision quest,"

headed for a midnight dip at the springs. Later that night, the desert gods dealt wind, dust, and morning drizzle.

The first morning's business, of course, was to hit the springs. Already, the other two playa planes had flown over to the road and parked at roadside by the springs. (The road is probably the safest year-round place to land.) The pilots, from western Oregon, had camped at the center of the flat. We soaked until the rain got heavy, then headed for the planes.

By the time I got to my plane, the surface had already become slippery. On departure, the goo was flying everywhere, coating even my prop. Now I know what it is like to drive a manure spreader – well almost. While circling for altitude, I noticed the pilots' tent partially submerged in the center of the flat. Apparently, they saw that they could not land in the water and abandoned their site until later.

- RW

OR - ASHL

Ashland

Ashland Muni – Sumner Parker Airport

Town	Ashland **ID** S03 (S-zero-3)
Coord	N-42-11.42, W-122-39.64
Elev	1894 ft, SE end high, 1.1%
Runway	12 - 30, 3603 x 75' asphalt
Freq	CTAF-122.8
Charts	Klamath Falls, L1, Lo2

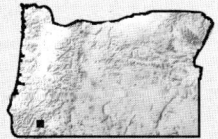

CAUTION: Airport information not for navigational use. Deer. Watch nearby hills and pole off Runway 12.

Photo by Jeremy Scott

> *Ashland is a charming theatrical hamlet, home of the Shakespeare Festival and other theaters. Close to outdoor recreation and historic Jacksonville. Theater is most active during non-winter months; skiing nearby in winter. $120-250*
> *CofC: 541-482-3486, www.ashlandchamber.com*

Airport Description

Ashland is nestled in a valley, the first Oregon airport off Interstate 5 just north of California. The Siskiyou mountains to the south provide a formidable barrier to VFR traffic when ceilings drop in winter. Once safely in the valley, landing at Ashland should be straightforward.

The runway is lighted dusk to dawn. When departing, preferred noise abatement procedures are to climb to 3000 on runway heading before turning.

Airport Services and Transportation

Airport phone: 541-482-7675. Fuel: 80, 100, and Jet. Skinner Aviation, 482-7675 provides a number of services: rental cars, aircraft repair, painting, and flight instruction.

Town center is located several miles from the airport, farther than most care to walk. Five motels are located next to Interstate 5, within a 20 minute walk. To get to this cluster of motels, walk southeast from the airport and turn right on Dead Indian Road. Walk 8 minutes and turn right on Highway 66. Walk northwest 12 minutes to the motels. Better yet, ask for a ride.

Most motels provide free shuttle service to the theaters. (Verify when making reservations.) Call 541-482-3065 for Yellow Cab. Budget has an office one mile from the airport, 800-527-0700.

Ashland **OR - ASHL**

Butler Ford, 541-482-2521, has a fleet of 50 cars that they rent for $30 and up. Skinner Aviation, 482-7675, handles Butler's cars at the airport. Make reservations in advance.

Lodging

Last-minute lodging is difficult to find, sometimes requiring a half-dozen phone calls to locate a room. Try calling the following motels. Those marked with "I5" are located at Interstate 5, one mile from the airport.

Features
B Full breakfast
C Continental
H Historic
J Jacuzzi
P Pets OK
R Restaurant
S Swim pool
T Transportation

Motel	Features	Price	Phone
Ashland Hills Inn	CJPST	$76-286	541-482-8310 / 800-547-4747
Ashland Motel - I5	PRS	$41-105	541-482-2561 / 800-460-8858
Bard's Inn - Best Western	CJPS	$82-130	541-482-0049 / 800-533-9627
Cedarwood Inn	JPS	$58-160	541-488-2000 / 800-547-4141
Columbia Hotel	H	$63-113	541-482-3726 / 800-718-2530
Best Western Windsor - I5	CJPST	$70-180	541-488-2330 / 800-334-2330
Shrew House B&B	BHJT	$102-135	541-482-9214 / 800-482-9214
Super 8 Motel - I5	P	$45-65	541-482-8887 / 800-800-8000

The Ashland Hills and Best Western Windsor appear to be the best conventional lodging in town. Ashland Hills Inn offers a 15% discount to pilots, April to October. Bard's is the closest motel to the theaters. Those that allow pets typically charge $6-10. Rich in history and short in modern-day luxury, the Columbia Hotel is located just one block from the theaters.

© 2001 by ALTA Research

The Ashland area has over three dozen B&Bs – far too many to mention here. Ashland's B&B Reservation Network, 800-944-0329, is a cooperative that can help find you a room. For area lodging and ticket reservations, call the Southern Oregon Reservation Center at 800-547-8052 or 541-488-1011.

Shakespeare Festival

In 1934, an enthusiastic young teacher named Angus Bowmer decided to produce for the city of Ashland a "Festival of Shakespeare." Convinced that Shakespeare would lose money, the city fathers insisted that they be allowed to stage boxing matches on stage by day to offset the Festival's losses. To their surprise, Bowmer quickly agreed. He reasoned that if Shakespeare could put up with bull and bear baiting between shows, he would not shun a contemporary equivalent.

Bowmer's project has come a long way since his first $400 production. The Festival's three theaters have earned themselves a worldwide reputation for technical excellence. Ashland has become the western proving grounds for top names in theater arts. The Festival now hosts over 300,000 enthusiasts per year.

Though known primarily for its 1200-seat outdoor Elizabethan Stagehouse, the festival supports two additional indoor theaters: the 600-seat Angus Bowmer and 1400-seat Black Swan. Performances for all three are likely to sell out during summer weekends. Call 541-482-4331 to reserve tickets as far in advance as possible. Tickets range from $21 to $50, and a backstage tour is $10. Their web site is *http://www.osfashland.org* .

If tickets are sold out, you have several alternatives. Get in line at the box office (near the Elizabethan Theater) somewhat before it opens at 9:30 AM. If luck is with you, you will get tickets; otherwise you receive a number. The number gives you a priority for purchasing returned tickets at a specified time later

Ashland **OR - ASHL**

in the day. Standing-room tickets for the Elizabethan Theater are also an option for purchase, in the morning or later in the day. Or, you may be able to purchase tickets (at standard price) from individuals just before show time.

Within blocks of the Festival, Ashland has grown several adventurous "off-Shakespeare" theaters. I had a delightful evening of dining and theater at the Oregon Cabaret Theater, a turn-of-the-century church that has been beautifully remodeled into a theater. Consider the following alternatives:

Actor's Workshop Theater	541-535-5250
Ashland Community Theater	541-482-7532
Oregon Cabaret, dinner theater	541-488-2902
Shakespere Festival	541-482-4331
S. Oregon Theater Arts	541-552-6348

Elizabethan and Angus Bowmer Theaters Photo by Hank Kranzler

Ashland Area

Ashland is an attractive little town, well worth a one- or two-hour stroll before dinner and theater. Start from the theater district and take time to browse **antique shops**, sample aromas from restaurants, or visit the 100-acre **Lithia Park** southwest of the Festival. Nearby are **golfing**, **swimming**, and **tennis**.

Past city limits, there's yet more to do within easy driving distance. **Hiking** and **skiing** are to be found in nearby mountains. Historic Jacksonville, 15 miles northwest, is a restored 1850s **gold rush town** that is worth an afternoon's visit. June through September, Jacksonville hosts the **Britt Festivals**, featuring **musical theater**, **jazz**, **dance**, **bluegrass**, and **classical**. Call 541-773-6077 for information.

Noah's World of Water offers **fishing** and **white-water rafting** adventures on the Rogue River. Choose a half-day or more of rafting on quiet or foaming waters. Prices range from $30 to $500. Call 541-488-2811 for information and reservations. Call the Adventure Center at 800-444-2819 for a broad spectrum of similar services. The Adventure Center claims to place a strong emphasis on covering every detail of the experience.

Downhill skiing is available Thanksgiving through April at Ski Ashland, 541-482-2897. Their four chair lifts give access to 22 runs with a vertical drop of 1150 feet. **Cross-country skiers** will find over 80 miles of trails within one hour of Ashland: Lake of the Woods, Fish Lake, Hyatt Lake, and Mount Ashland.

See **OR-PINE** for additional area information.

- RW

© 2001 by ALTA Research

OR - BAND

Bandon

Bandon State Airport

Town	Bandon **ID** S05 (S-zero-5)
Coord	N-43-05.24, W-124-24.57
Elev	119 feet
Runway	16 - 34, 3600 x 60' asphalt
Freq	CTAF-122.8, Lights-3x5x7x
Charts	Klamath Falls sectional

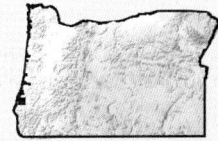

CAUTION: *Airport information not for navigation. Trees at each end. Coastal winds and weather.*

> *Bandon, a pleasant coastal community, with reconstructed Old Town, scenic coastal topography, and good restaurants. Perfect for an overnight or weekend visit. $7-200 CofC: 541-347-9616, www.bandon.com*

Airport Description

Bandon is a typical coastal airport. Located two miles southeast of town and within a mile of the ocean, it is easy to spot, once the clouds have cleared. The region is typically overcast in the morning; expect burn-off by 10:00 to noon in summer.

The regional **coastline** is beautiful. If you arrive from the south, you fly over Cape Blanco, the most western point in the lower 48. At its 1870 **lighthouse**, winds have been clocked at over 180 mph. If you arrive from the north, plan your route to follow the coast at low altitude. From the north, between Florence and North Bend, you fly over at least twenty miles of **coastal dunes** covered with ATVs speeding around like little ants. Don't let the fascinating view prevent you from keeping an eye out for other low-flying tourists.

Services and Transportation

Airport phone: Frank's, 541-347-2022. Fuel: 100LL. Frank's Flight Service provides friendly service and maintenance. Frank sometimes has a courtesy car available for quick runs into town. Junkers for rent at $25 per day or $15 for a half day.

Bandon has unconventional transportation services. Try connecting with Rides by Donation by calling 541-530-6663. Expect to pay $7 for the ride into town. County sponsored Dial-A-

Bandon **OR - BAND**

Ride provides rides for $2 per person between 9:00 and 16:00 on weekdays, 347-4131.

To get to town, drive 0.2 miles west to Highway 101. Turn right, and drive 2 miles north to the edge of town. At 2.7 miles you approach Old Town, which is clearly marked with an overhead sign.

Lodging

Bandon offers a variety of lodging, both inside of town and along its beautiful coast. Lodgings marked with an asterisk are within reasonable walking distance of Old Bandon.

Features
B Full breakfast
C Continental
H Historic
J Jacuzzi
P Pets OK
R Restaurant
S Swim pool
T Transportation

Lodging	Features	Price	Phone
Bandon Beach Motel	JPS	$48-58	541-347-4430
Bandon Dunes golf resort	JRT	$65-850	888-345-6008
			541-347-4380
Gorman Motel *	CJPST	$58-95	541-347-9451
Harbor View Hotel *	CJT	$50-135	800-526-0209
			541-347-4417
Inn at Face Rock	CJPRST	$63-240	800-638-3092
			541-347-9441
Lighthouse B&B *	BJT	$117-200	541-347-9316
Sea Star Guesthouse *	RT	$48-101	541-347-9632
Sunset Motel	JPST	$55-175	800-842-2407
			541-347-2453
Windermere Motel	H	$70-145	541-347-3710

Five miles north of town, the new Bandon Dunes golf resort has modern lodging and two excellent **golf courses**. One of the courses is **world class**, and said by golf aficionados to be the best course to be built in the U.S. in years.

The best deal sits in the heart of Old Bandon – the Sea Star Guesthouse. Don't be put off by their low $16 price for a single

They offer low-priced "hostel" accommodations, as well as the standard fare. The rooms are clean and attractive.

The Harbor View Hotel overlooks Old Bandon and the harbor. The Lighthouse B&B is just west of Old Bandon at a dramatic location on the Coquille River. The Gorman Motel is further southwest, with a hot tub and view of the beach.

The following motels are located on the Beach Loop Drive, south of town. They all have great views. The Bandon Beach Motel is old and lacks class, but is reasonably clean. Next in line at two miles is the Sunset Motel, a nice place. Drive another half-mile to the Windermere Motel, a small group of cottages at a special beach location. Across the street is the Inn at Face Rock with nearby **golf course** and dining facilities.

Restaurants

For a quick **gourmet snack**, you can order an inexpensive seafood cocktail at the Bandon Fisheries Market at First and Chicago. If you've got the munchies, you'll have no problem finding a food-fix in Old Town.

Bandon has over two dozen restaurants, many of which are good-to-excellent. On weekends, it is both prudent and worth the effort to make reservations in advance at the **better restaurants**. Harp's, 541-347-9057, has excellent halibut, house pasta, and a filet mignon that is marinated in garlic and teriyaki. The "house" sweet onion soup is a tasty creation.

Billy's Smoothboars has a sports motif, and is the place for tasty ribs and roast beef, 347-2373. You get the best deal if you eat in the lounge.

For **seafood** cuisine with a harbor view, try the two-story Boatworks, 541-347-2111, near the south jetty on the Coquille River.

© 2001 by ALTA Research

Bandon **OR - BAND**

Things to Do and See

Take time to walk around Old Bandon. I think you will agree that the local folks have tastefully renovated the nucleus of their oceanside community. You'll find a number of **shops to browse**. I enjoyed both the Winter River Books and Gallery and the Second Street Gallery, both on Second Street. The Second Street Gallery is the largest **gallery of fine art** between Seattle and San Francisco.

After passing stores like the Cranberry Sweets shop, I made the connection between **cranberries** and the curious ponds you might have noticed on your approach to the airport. No, those ponds are not for sewer treatment. (Bandon's waste is treated by a complex of concrete tanks covered with whales and dolphins at the northeast edge of Old Bandon.) One thousand acres of such ponds, or bogs, are used for growing cranberries. For the visitor,

Bandon sewer treatment plant, a must-see work of art

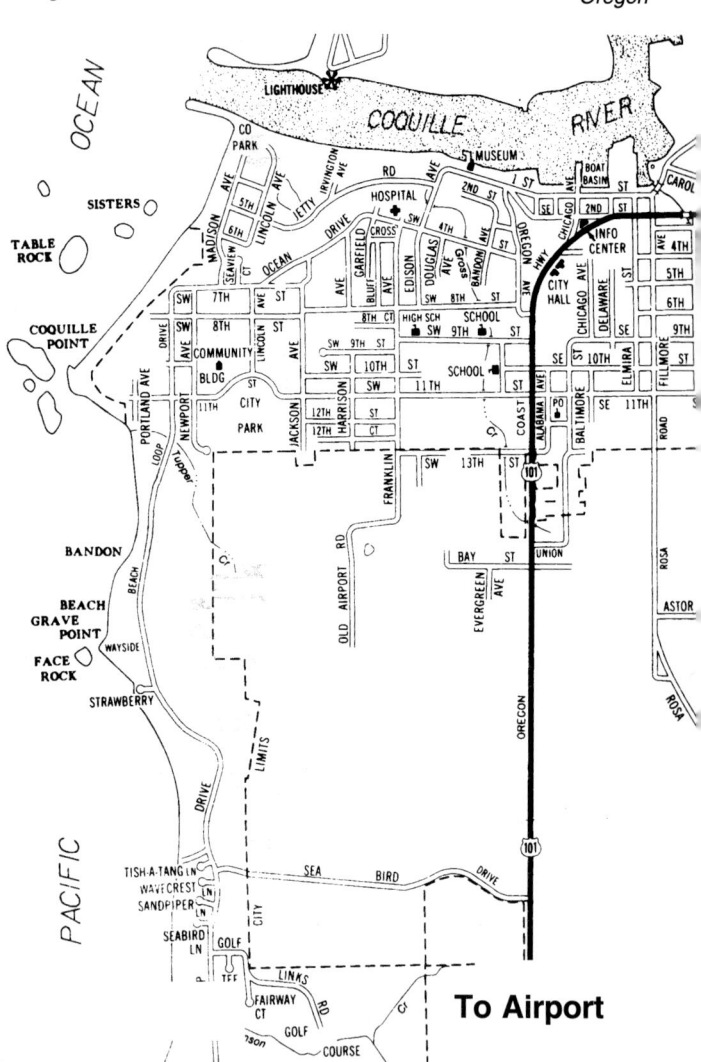

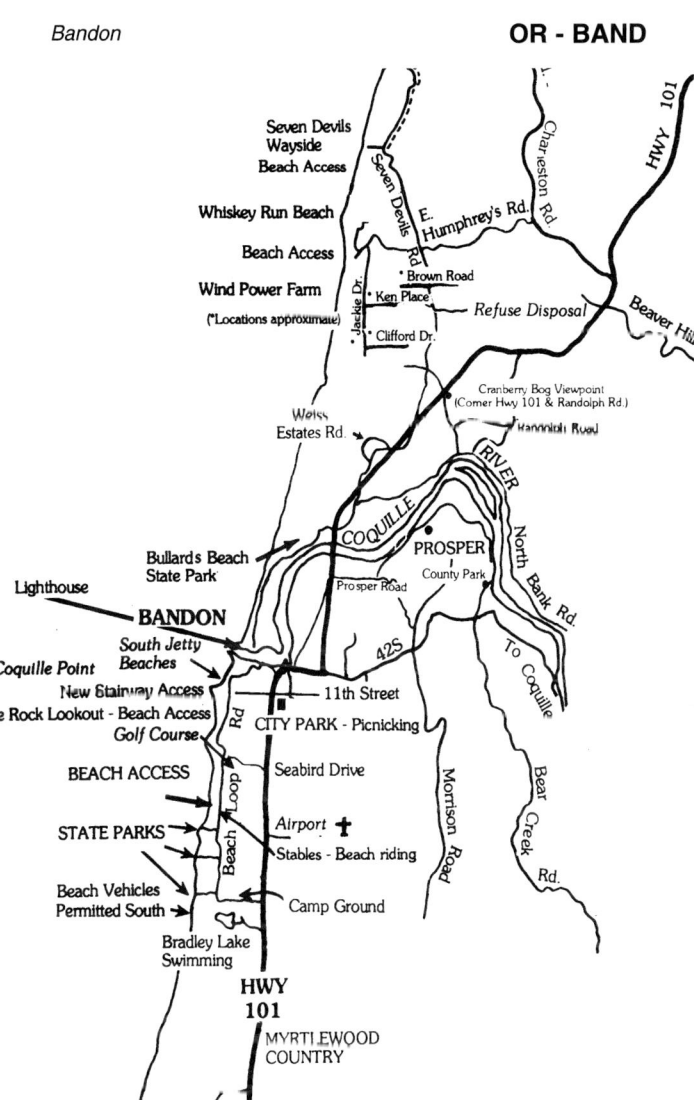

this means shops full of cranberry treats, and a four-day September Cranberry Festival.

You can extend your stroll by visiting **Bullards Beach State Park**, one mile north of Bandon off Highway 101. Explore the park's 1,266 acres with hiking trails that lead to driftwood-cluttered beaches. Across the dunes, you view the **1896 Coquille River Lighthouse** – a beauty to behold, but not open to the public.

The new **Bandon Dunes golf resort** is five miles north of town, 888-345-6008. It offers premier golfing experiences at its two courses: Bandon Dunes and Pacific Dunes.

Given time and transportation, consider enjoying the beautiful Oregon coast from ground level. Head south on the **Beach Loop Road**, which exits Old Bandon at its southwest corner.

Visitors browse the shops, while morning sun works to clear the sky

After a mile, you have access to **coastline bluffs** and a lovely stretch of **beach**. After two-plus miles from town, and you pass the Inn at Face Rock **golf** resort. Nearby Face Rock is said to be the face of an Indian princess, turned to stone by an evil spirit.

Bandon Beach Riding Stables, 541-347-3423, is located at 2640 on the Beach Loop Drive. They provide beach rides to people of all ages. Further along the loop, you pass various state parks before the road turns east, passes a campground, and connects to Highway 101. The airport is north of this intersection.

Surprise the kids with a visit to the 26-acre **West Coast Game Park**, seven miles south of Bandon off Highway 101. On 26-acres of wooded grounds, 75 domestic and **exotic animal species** make their home, including snow leopards, lions, tigers, zebra, cougars, bears, camels, antelope, bison, monkeys, and peacocks – 450 birds and animals in all. The owners do exten-

Second Street Gallery, a center for fine arts

sive breeding and claim to have the country's largest wild animal **petting zoo** with baby animals. Entry fee is $9 for adults, and less for children and seniors. Open March through November, call 541-347-3106 for information. Visit late in the day when the crowds are down and the cats are active.

Some twenty miles south is **Cape Blanco State Park**, 541-332-6774. The park covers most of the cape and includes historic buildings and artifacts from Hughes Ranch, dating back to the end of the 1800s. As mentioned above, Cape Blanco and lighthouse are at the most western tip of the 48 states. Twenty miles north of Bandon, you can visit **Shore Acres State Park**, which houses a worthy **arboretum** and is the gateway to **Simpson's Beach**.

Another interesting place to visit is the **Bandon Glass and Art Studio,** 541-347-4723. The studio is now conveniently located in town at 240 Highway 101. Just north of town, you can see cheese made by hand at **Bandon's Cheddar Cheese Factory**, 800-548-8961.

Events

Bandon has a number of events throughout the year. Every Memorial Day weekend, they have their **Stormwatchers' Seafood and Wine Festival**, followed by the **Sandcastle building contest** on Monday. The Coquille Indian Tribe does their **salmon bake** on the weekend prior to July 4th. On the **4th of July**, the town celebrates with more fish, fiddle music, and fireworks. The Cranberry Festival occurs in September.

This is a small town of less than three thousand souls. The festivals are correspondingly modest. Check with the Chamber of Commerce for dates and details, 541-347-9616.

- RW

© 2001 by ALTA Research

OR - DIAM

Diamond Peak Wilderness

Crescent Lake State Airport

Town	Crescent Lake Junction **ID** 5S2
Coord	N-43-31.96, W-121-57.00
Elev	4810 feet, NW end high
Runway	13 - 31, 3760 x 125'
	SE end paved, NW end gravel
Freq	CTAF-122.9
Chart	Klamath Falls sectional

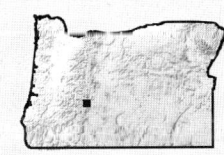

CAUTION: Airport information not for navigational use. Call 503-378-4880 before using. Unusual drafts caused by crosswinds. Trees surrounding the runway. Watch density altitude.

> *Diamond Peak Wilderness offers laid-back mountain camping. Wilderness campsites are off the beaten track and relatively easy to reach from the airstrip. $0*

Airport Description

VOR reception drops out at approach altitudes, but the airstrip is clearly visible and easy to locate. While in the pattern for landing, note the useful escape route at the northwest end. The runway conveniently aligns with the railroad tracks, providing an extra mile almost free of obstructions.

The airport is closed during winter months. A local resident claims the airport is usable from May until mid-September.

Services and Transportation

Phone: Oregon Aeronautics, 503-378-4880 (not at airport). Call this number for information before landing. Tiedown is free. Phones, restaurants, and a small grocery store are located one block from the tiedown area. Camping adjacent to the strip appears to be feasible.

Conventional transportation is not available and not required. Trailheads are less than two miles from the tiedown area.

Overview

The Diamond Peak Wilderness Area offers hikers a number of advantages over better-known destinations: **Trailheads** are

Diamond Peak Wilderness **OR - DIAM**

close to the airstrip. **Swimming** is available. By California standards, the area is relatively unused. Minimal climbing is required since Fawn Lake's elevation is only 850 feet above the airport. Trails are pleasant and easy to follow. Camping permits are not required. The only real disadvantages are mosquitoes and shortage of frequent alpine views.

Unfortunately, the strip is closed during the winter months. Year round availability would provide access to an extensive cross-country ski area that is maintained just north of the airport.

Fawn Lake Loop

The Fawn Lake Loop can be a comfortable one-nighter, or it works well as a full weekend experience. Within two and one-half hours of landing, you can hike to a campsite on Fawn Lake. Topo maps for the area are *Odell Lake* and *Waldo Lake*. Though not necessary, they come in handy. The trail is smooth, and light hiking boots or jogging shoes work out well. A compass is recommended; insect repellent is essential.

To find the trailhead, walk to the northwest end of the runway. Intersect and follow the railroad tracks in exactly the same direction. At the point at which the tracks begin to bend to the left, carefully watch for the trailhead on the left. It is located an eighth of a mile south of the tip of Odell Lake, and is easily identified by the blue, diamond-shaped trail markers. Walking time from tiedown to trailhead is 35 minutes. An additional hike of less than two hours is required to reach Fawn Lake.

The trail joins Fawn Lake at its northeast shore. Follow the trail clockwise to reach good campsites at the southern and western shores of the lake. A scenic site is located on the

© 2001 by ALTA Research

inlet near the tip of the peninsula on the western edge of the lake. At most sites, the lake's sandy bottom and warm temperature provide refreshing **shallow-water swimming**.

Two other lakes are easily accessible from the point at which Odell Lake Trail joins the lake at its northeast shore. From this reference point, Stagg Lake is about one and one-half miles, and Saddle Lake is two miles. Follow the trail west.

After about twenty minutes, you officially enter the Diamond Peak Wilderness Area. A detailed map is posted below the wilderness area's sign. Seven hundred feet past the sign, a trail to Stagg Lake branches off to the right. If you follow the branch to Stagg Lake, you arrive in about ten minutes. The lake is nice to look at, but not much for swimming.

Continue 20 minutes from the wilderness area's boundary sign along the established trail to Saddle Lake. This lake is about a third the size of Fawn Lake, contains some good campsites, and appears to be suitable for **shallow swimming**. Be forewarned that the *Waldo Lake* topo map has inaccurate representation of trails in this general area.

When returning to the airport, you may return via the familiar Odell Lake Trail, or opt for the Crescent Lake/Town Trail. The latter route eventually intersects the railroad tracks at the town of Crescent Lake. From this point, follow the road two miles northeast until you reach Runway 31 at Crescent Lake Junction.

Deeper Penetration

Serious backpackers note: The **Pacific Crest Trail** is fourteen miles by trail west of the airport.

- RW

OR - FLYM

Flying M Ranch

Flying M Ranch Airport

Town	Yamhill **ID** OR05 (oh-R-zero-5)
Coord	N-45-21.72, W-123-21.34
Elev	448 feet, W end high, 1%
Runway	7 - 25, 2135 x 40' dirt
Freq	CTAF-122.9
Charts	Seattle sectional

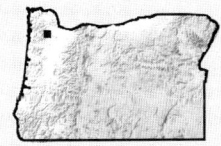

CAUTION: *Airport information not for navigational use. Last-minute go-around dangerous due to trees and terrain at W end. Experienced dirt-strip pilots only. No night operations. Ultralights.*

> *The Flying M Ranch offers riding, pack trips, hay rides, hiking, swimming, and court games. The Flying M's beautiful log lodge is surrounded by hillside forests and pastures. $10-250 Lodge: 503-662-3222. www.flying-m-ranch.com*

Airport Description

The ranch's dirt landing strip is located 16 miles on the 251 radial from Newburg's UBG VOR. The runway, which seems narrower than the listed 40 feet, has trees close by. Look for the lodge at the west end on the northern side of the strip.

Land to the west and depart to the east. Announce intentions on 122.9. Make go-around decisions early, as a last-minute decision could put you in the trees. The ranch advises against tailwind operations, and night operations are not recommended.

Services

Airport phone: 503-662-3222. Fuel: none. The Flying M Ranch is a recreational facility that provides hearty meals, lodging, camping, and entertainment.

Flying M Ranch

Over 100 years ago, the Traveler's Home served as an overnight resting place for travelers on the Trask Mountain stage coach route. Today the site is home for the Mitchell family and their Flying M Ranch. They built their first lodge in 1971, but it burned to the ground in 1983.

The Mitchells completed rebuilding their beautiful replacement in 1985. The lodge is a massive structure built of enormous Oregon logs. The facility includes an Olympic-size bar and two large dining rooms separated by a prominent stone fireplace. The front entrance is festooned with antique chain saws, while inside walls and ceilings are decorated with hunting trophies and historical artifacts. Entrees at the lodge range from $6 to $20.

The 650 acre ranch is a unique fly-in experience. Flying the approach and landing may cause anxiety for the pilot; but once on the ground, visitors can look forward to a laid-back vacation. The grounds sustain Oregon green vitality through summer

months. Winds leisurely wag tall firs that look over the valley. Mountains rise above horses pasturing on a nearby knoll. Lush vegetation surrounds the stream that passes through the campgrounds and past the lodge.

Rooms at the Flying M Motel cost from $60 to $70. Cabin rental costs from $85 to $200. Overnight camping costs $12. Most campsites are near the river and several have adjacent airplane parking. Camping sites are primitive, but toilets and showers are available within walking distance. Call 503-662-3222 for reservations, 8:00 to 5:00 on Monday through Friday.

The ranch offers a full spectrum of dude ranch activities. **Horseback riding** and **wagon rides** are everyday activities. The Flying M **swimming pond** provides a refreshing reward after several hours in the saddle. Trail rides generally cost $15-$17 per hour. Near sundown, try their **steak fry ride**, which costs $40-45 for adults and $22 for children. On certain days, the Flying M runs an **overnight trail ride** that costs $175 per rider.

On-site activities include **tennis**, **volleyball**, **horseshoes**, **basketball**, and **hiking**. The lodge features **live country western music** on Friday and Saturday nights. Besides the daily and weekly activities mentioned above, the ranch lists over three dozen special events each year. A sampling includes: **black powder shoots**, special **brunches**, **workshops**, and **holiday celebrations**.

Large facilities are available for your own fly-in, meeting, retreat, or seminar. Companies like Hewlett-Packard have been known to exploit the Flying M's ambiance and isolation for off-site meetings. The Flying M is even available for purchase. Call 503-662-3222 for information.

© 2001 by ALTA Research

- RW

OR - HOOD

HOOD RIVER

Hood River Airport

Town	Hood River **ID** 4S2
Coord	N-45-40.36, W-121-32.19
Elev	631 feet, W end high
Runway	7 - 25, 3040 x 75' asphalt
Freq	CTAF-122.8, Lights-2x5x7x
Charts	Seattle sectional

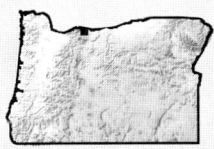

CAUTION: *Airport information not for navigational use. Gliders and crop dusters. Vehicles at W end. Pole near Runway 7.*

> *Hood River, at the eastern end of the Columbia Gorge, a windsurfer's haven. Hood River Railroad rides, boating, scenic drives, museums, restaurants, and breweries. $100-400*
> *CofC: 800-366-3530, www.hoodriver.org*

Airport

Hood River and its airport are just south of the Columbia River, and 20 miles north of Mount Hood. The airport is at the eastern end of Columbia Gorge. Local pilots are well aware that The Gorge has major influence over local weather. Gorge weather is a mixed blessing for pilots. Often, it provides the only VFR route east and west past the Cascade Mountains. On the other hand, its winds can be brutal, and visibility and ceiling can change rapidly.

The terrain surrounding the airport is friendly. Winds typically favor Runway 25, which is the default runway.

Services, Transportation, and Lodging

Airport phone: 541-386-1133. Fuel: 100 octane. Flightline Services has a lounge, pilot supplies, and a courtesy car. The Hood River Taxi is 541-386-2255. The following rental agencies are located in town, and not at the airport. Connections with the airport require a taxi or advance arrangements, and procedures become more complex on weekends.

Columbia Scooters	541-386-5220
D&S Reliable Rentals	541-386-4039
Enterprise (at The Dalles)	541-386-9480
Hood River Dodge	541-386-3011

Hood River **OR - HOOD**

Located near the train station, the Hood River Hotel is near the center of town with grand old rooms and a quality Italian restaurant. The Best Western Hood River Inn has scenic views of the river. The Columbia Gorge Hotel is the finest and most expensive. Inns marked with an asterisk are within easy walking distance of the train station.

Features
B Full breakfast
C Continental
H Historic
J Jacuzzi
P Pets OK
R Restaurant
S Swim pool
T Transportation

Beryl House B&B	BHPT	$82-87	541-007-5567
Columbia Gorge Hotel	BHPR	$165-300	800-345-1921
Hood River Hotel *	CHPR	$53-120	800-386-1859
Hood River Inn B.W	JPSR	$70-195	800-828-7873
Inn at the Gorge B&B *	BH	$90-114	541-386-1129
Oak Street Hotel *	HT	$54-59	541-386-3845
Vagabond Lodge	P	$43-83	541-386-2992

Hood River Area

Hood River is one of a number of interesting towns that populate the shores of the Columbia River. Located 50 miles east of Portland, Hood River has become a center for sports people who crave wind and water. For years, windsurfers have thrilled observers as they ride the westerly blast from The Gorge. More recently, kiteboarders have been stealing the show.

Long before recorded history, The Gorge area served a thriving Native American trade network that stretched far beyond the Pacific Northwest. Just 200 years ago, Lewis and Clark brought this region to the attention of the white man. By 1850, pioneers arrived by the Oregon Trail to settle by the river. Hood River was settled in 1854, and incorporated in 1895.

Early settlers quickly built portages, canals, and eventually locks to circumvent the rapids. Railroads arrived to carry away

© 2001 by ALTA Research

riches in timber and to provide transportation to the coast. Fishing and farming became top industries. The farmers learned to use the Columbia for irrigation, and of course dams were erected to siphon energy from the mighty Columbia River. The nearby Bonneville Dam was completed in 1938.

A scenic segment of historic railroad track between Hood River and Parkdale has been restored. Originally used to transport timber and fruit, the privately owned Hood River Railroad now caters to visitors. The leisurely ride offers snowcapped views of 11,245-foot Mt. Hood. While the views were nice, the good natures of the train personnel and the quality of the food left me with a smile. Oh, I guess the Champagne helped too.

Although the Brunch Train ride is the most popular, consider the options: Dinner Train, Murder Mystery Dinner Train, Easter Egg Train, Western Train Robbery and Country Barbecue, Hal-

The Hood River Railroad train has a full-service bar and kitchen car

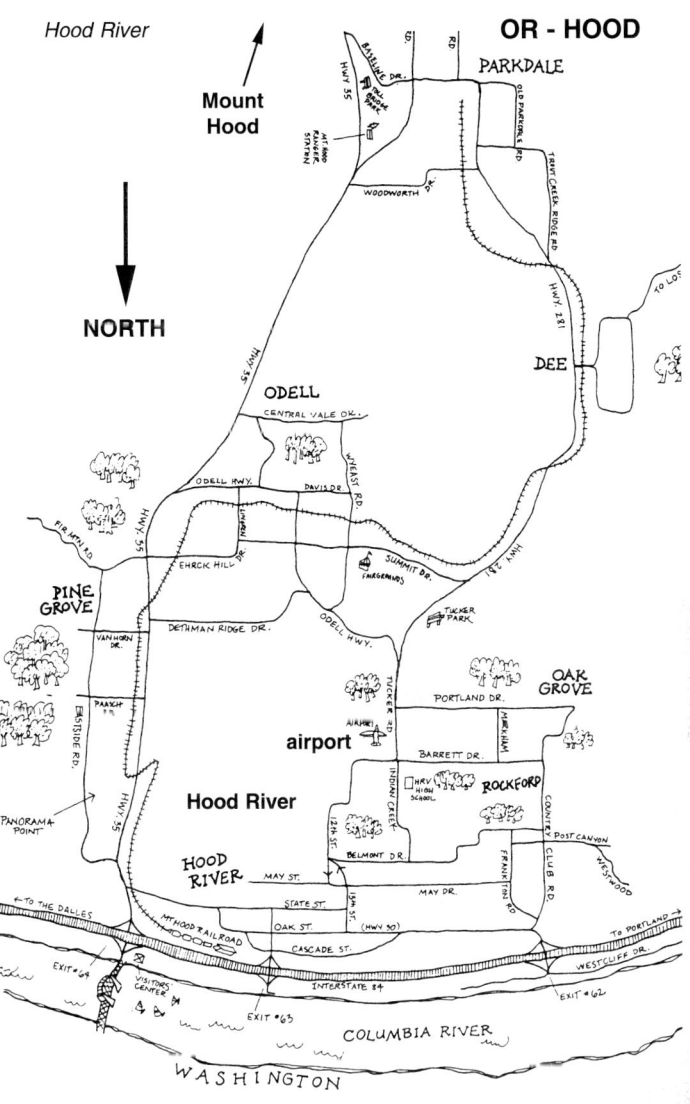

Parkdale Station, end of the line

Hood River **OR - HOOD**

loween Spook Train, Circus Train, and more. The excursion trains cost $25 for adults. Dinner and Brunch trains cost $70-80. For details, call 800-872-4661 and see *www.mthoodrr.com* .

The **Jesse & Winifred Hutson Museum**, 541-352-6808, sits by Parkdale train station. The museum includes local historic items, including personal mementos, native artifacts, pioneer hand tools, and heavy equipment. In Hood River, you can visit the **Hood River County Historical Museum**, 386-6772.

The Sternwheeler Columbia Gorge offers inexpensive 2-hour **boat cruises**, Champagne Brunch Cruises, and Dinner Dance Cruises with live music. The boat departs from Cascade Locks, a 20 minute drive from Hood River. Rides cost $15-50 and less for children. Call 800-643-1354 or see *www.sternwheeler.com* .

The **Historic Columbia River Highway** is an early example of how the impossible can be achieved without the benefit of computers. Inspired by Sam Hill and engineer Samuel Lancaster, this elegant paved road was built between The Dalles and Portland in 1913-1922. Although it has been replaced by Interstate 84, many of its scenic segments have been restored for the enjoyment of visitors. The Hood River Chamber has maps and literature that can guide you to vistas, falls, and other points of interest.

In town, windsurfers will want to browse the shops for the latest in wind and water gear. Windsurfer or not, make a point to watch the afternoon **windsurfers** and **kitesurfers** within walking distance at the river. Fascinating.

Rather than being pulled by a board-mounted sail, kitesurf-

ers are pulled by an airborne parasail. Kitesurfers can perform airborne stunts that surpass the norm for traditional windsurfers. Further, kitesurfers need only 8-10 mph for jumping 10-15 feet into the air, while wind surfers need 20-30 mph just to get started. Like paramotor and parasail gear, kitesurfer equipment can be easily stowed in a small package.

Surf photos: Stephanie Mazzara   Kitesurfer: Tony Barbʳ

Hood River is a great place to **learn windsurfing or kitesurfing**. The local shops are pleased to rent or sell gear to visitors. The following schools can get you started: Big Winds

Hood River **OR - HOOD**

Hood River, 888-509-4210; and Rhonda Smith Windsurfing Center, 541-386-9463. Expect to spend about $150 per day for several hours of instruction and rentals. OutdoorPlay, 877-72-KAYAK, teaches **kayaking**.

I enjoyed visiting the **International Museum of Carousel Art**, located on Oak Street in the heart of town. The museum is home to the largest and most comprehensive collection of antique carousel animals. I loved the sound of their player band organ, a Whurletzer 153. The entry fee is $5, a good value if you have interest in lost art such as this. Information: 541-387-4622 or *www.carouselmuseum.com* .

Hood River has two breweries near the center of town. The **Full Sail Brewery** has earned national acclaim for its microbrew-style beers. You can tour the brewery and sample a tray of beers in their pub at no charge.

The **Big Horse Brew Pub** is another required stop for beer lovers. Local wine tasting occurs at **Flerchinger Vineyards**, 800-516-8710; and, **Hood River Vineyards**, 541-386-3772.

With over two dozen diverse **restaurants** from which to choose, chances are, you will find the right place to dine. A good place to start is the center of town near Oak and Cascade Streets. The Hood River Hotel's Pasquale's Restaurante Italian serves excellent meals. At the Skylight Theater, you can order pizza, flop in a sofa, and eat while watching a film.

Hood River has two 18-hole **golf courses**: Hood River Golf & Country Club, 541-386-3009; and Indian Creek Golf Course, 386-7770. **Horseback Riding in the Gorge** provides **guided horseback tours** near Cascade Locks, 374-8592.

© 2001 by ALTA Research

If you have never **paraglided**, here's your chance. Call 541-387-2112 to arrange for a 10-60 minute tandem flight for $150. Summer thermals can keep riders aloft for up to two hours.

Gorge **Fly Fishing** Expeditions, 541-354-2286, offers 1-3 day floats down the Deschutes River. Gorge Fly Shop, 386-6977, conducts training sessions and excursions to local lakes and streams. Costs run about $150 per day. Phoenix Pharms has a **you-catch trout** lake for family fun, 352-6090.

Hood River is well positioned for **automobile excursions**. Consider visiting any of the following: Mt. Hood's Timberline Lodge, Bonneville Dam, Cascade Locks, Stonehinge at Maryhill, The Dalles Dam and its free tour train, the Fort Dallas Museum, and of course scenic drives on the historic highway. A number of majestic water falls can be found along the highway west of Hood River.

Big Horse Brew Pub near town center

OR - HORS

Horse Ranch

Horse Ranch USFS Airport

Town	ref. La Gande
ID	60R9 (6-zero-R9)
Coord	N-45-14.8, W-117-37.5
Elev	3600 feet
Runway	9-27, 2800x100' turf, gravel
Freq	CTAF-122.9
Charts	Seattle sectional

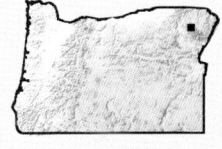

CAUTION: *Airport information not for navigational use. Difficult mountain strip. Mountains close by. Unpredictable SW wind. No go-around for some aircraft. Traffic from Minam strip. Ditches on W side. Mud spots in spring. Tall grass possible. Elk. Horse/hiker trail crosses runway. No radio. Not officially maintained.*

> *The Horse Ranch, historically a wilderness dude ranch, now owned by the USFS. Uncertain future. Alternate landing strip for the Minam Lodge. Access to trailheads. $0 Enterprise USFS: 541-426-4978*

Airport Description

Look for two grass strips in the Minam River valley. The southeastern strip is the Horse Ranch runway. Mountains rise steeply on either side. Circle once before landing and announce intentions on 122.9. Setup the approach as though you were going to land at the northwestern strip (located one mile northwest). A go-around may not be possible for some aircraft. Consider approaching low and adding power to stretch your glide to the Horse Ranch strip. Follow the dog-legs between trees so you can get low enough to avoid a go-around. Standard departure is downstream to the north.

Although the runway is not officially maintained, the USFS, Blue Mountain Backcountry Horsemen, Oregon Pilot's Association, and other groups do their part to keep it safe. The Horsemen purchased a mule-pulled mower to keep the grass down. Some pilots use this runway as an alternative to the strip at Minam Lodge, **OR-MINA,** because they believe it to be less hazardous.

Services, Transportation, & Lodging °

Phone: USFS, 541-426-4978 (not at airport). No fuel, phone, nor maintenance services. The lodge and cabins are no longer available for guests; they are for the caretakers. Registration is mandatory for all visitors. The permit box is 100 feet north of the windsock.

© 2001 by ALTA Research

Horse Ranch (wilderness) **OR - HORS**

There is no transportation, but the walk to Minam Lodge takes only 15 minutes. Cross the runway-meadow to the west, and follow the trail north through the trees to Minam Lodge.

The Horse Ranch

Prior to recent purchase by the government, The Horse Ranch and Minam Lodge were the only commercial fly-in wilderness retreats of their kind in Oregon. There are no roads to the ranch and it is accessible only by foot, horseback, or airplane. The site was first operated as a pack station in 1931. Over the years it evolved into a dude ranch as it passed through the hands of several owners. While in private hands, capacity was 25-30 guests.

The ranch sits in a valley in the northwest quadrant of the Eagle Cap Wilderness, the largest designated wilderness in Oregon. The area contains 58 lakes and a small ice field; its peaks

range to 9845 feet. There is **fishing** in most lakes and streams and the area is rich in flora, birds, and wildlife.

The Horse Ranch fits right into this rugged and beautiful environment. Its sturdy cabins were built with timber from the ranch. The cabins feature fire places and hot showers. They sit by the Minam River, which flows down from high lakes and has good trout fishing. See **OR-MINA** for other local information, including map and trails.

The Forest Service asks that visitors be sensitive to the wilderness environment. Party size is limited to 12 people. Exercise courtesy for riders and hikers near the runway. For noise abatement, use a straight-out departure, rather than circling overhead.

The site is being evaluated for future use. If you would like to do your part to keep wilderness airports open and available for use, send comments to Ms. Kendal Clark, U.S. Forest Service, 88401 Highway 82, Enterprise, OR 97828. The USFS deserves encouragement when it maintains diversity in the wilderness.

OR - MCMI

McMinnville

McMinnville Airport

Town	McMinnville **ID** MMV
Coord	N-45-11.67, W-123-08.16
Elev	163 feet
Runway	4 - 22, 5420 x 150' asphalt
	17 - 35, 4676 x 150' asphalt
Freq	CTAF-123.0, Lights-3x5x7x
	ASOS-135.675, ILS-110.9
Charts	Seattle sectional

CAUTION: Airport information not for navigational use. Helicopters and crop dusters.

> McMinnville, an attractive Oregon town with interesting lodging, restaurants, and nearby vineyards. Evergreen Aviation's striking new air museum is home to the Spruce Goose. $100-200
> CofC: 503-472-6196, www.mcminnville.org

McMinnville Airport

McMinnville Airport is a modern dual-runway airport with instrument approaches and flight service station. Located 30 miles southwest of Portland in the Willamette Valley, arrival and departure are straightforward. Oregon's only flight service station is on field, but landing advisories are handled by Cirrus Aviation.

Services, Transportation, and Lodging

Airport phones: 503-434-7411; Cirrus Aviation, 472-0558; Northwest Airmotive, 472-4790. Fuel: 100 and Jet. The flight service station is a short walk from the FBO. With friendly staff, the FSS is worth a visit. Their web address is *www.faa.gov/ats/mmvafss* . Glider rides are available from Cascade Soaring, 472-8805. North American T6 rides are offered by Gary/Warbird Rides, 709-3059.

Cirrus Aviation rents their courtesy car for $5 for several hours or $20 for overnight. Their car is for local use only. Enterprise, 503-472-4010, can pickup from the airport, but is closed Saturday afternoons and Sunday. West Valley Taxi, 503-472-1714, provides service from airport to town center for about $5. The Evergreen Aviation Museum is within easy walking distance of the airport.

© 2001 by ALTA Research

McMinnville **OR - MCMI**

McMenamins Hotel Oregon has the best location, the best food and beer, the most interesting decor, and the most art of all lodging in McMinnville. The hotel has 42 rooms on the upper 3 stories, with restaurant and bars at the basement, ground floor, and roof. Breakfast is included with lodging. Because the hotel is typically booked on weekends, reservations are a must. Rooms are available with and without private bath. The McMenamins brothers have created over 30 exceptional bars, restaurants, and inns throughout Oregon. See **OR-TROU** for a full description of their work.

Features
- **B** Full breakfast
- **C** Continental
- **H** Historic
- **J** Jacuzzi
- **P** Pets OK
- **R** Restaurant
- **S** Swim pool
- **T** Transportation

McMenamins Hotel Oregon	BHR	$75-125	503-472-8427
Motel Paragon	CST	$42-49	503-472-9493
Safari Motel	R	$60-64	503-472-5187
Steiger House B&B	BT	$70-130	503-472-0821
Vineyard Inn B.W.	CJS	$72-117	503-472-4900

Lunch on the roof at McMenamins Hotel Oregon

McMinnville is a pleasant town. Its city streets are lined with trees and historic buildings. Third Street is the main promenade through town. Nick's, my favorite Italian restaurant in Oregon, sits next to McMenamins on Third Street. At Nick's, you can enjoy a leisurely 5-course meal, guaranteed to leave the tummy warm and satisfied. The meal is worth every penny of its $35 fee.

McMenamins is well known for their excellent microbrew, which is created at other sites. Several blocks from McMenamins, you can sample McMinnville local microbrew at the Golden Valley Brew Pub. McMinnville is surrounded by Yamhill's 50 wineries, many of which have tasting rooms.

Evergreen Aviation Museum

The Captain Michael King Smith Evergreen Aviation Museum is an impressive 121,000 sq-ft structure that houses the

Brothers examine the radial engine on a military DC-3

McMinnville **OR - MCMI**

Spruce Goose and other vintage military and civilian aircraft. Although Evergreen's primary business is international leading-edge helicopter operations, they had the soul to divert profits to a fascinating corner of aviation history. In 1962, Evergreen purchased the Hughes Flying Boat. In 2001, they opened their museum and the "Spruce Goose" to the public.

The Hughes Flying Boat was designed to fly troops and cargo over enemy submarines, which had been brutalizing American supply lines during World War II. It was to be the biggest airplane ever built. Howard Hughes took on the monumental project at a point when time and materials were in short supply.

His team created an 8-engine floating airplane with wingspan of 320 feet and length of 219 feet. The entire airframe and surface structures were built of laminated wood. All primary control surfaces except the flaps were fabric covered. While

Hughes Flying Boat

Photo: Evergreen Aviation Museum

Dimensions:
Wingspan: 319.92' (97.54m)
Fuselage: 219' (66.75m)
Tailspan: 113.5'
Vertical Tailspan: 49.5'

Gross Weight: Approx. 400,000 lbs. (181,440 kg.)

Fuselage Height: Approx. 30' (9.14m.)
Overall Height (at tail): approx 80'

Hughes met the primary objective of designing an airplane of mammoth capacity with minimal need for dwindling war materials, the flying boat became a political football. Indeed one U.S. Senator dubbed the project a "flying lumberyard."

No doubt such ridicule drove Howard Hughes to end the project in a manner that history would not forget. On November 2, 1947, Howard Hughes and crew fired up the Goose for a taxi test. Thrilling thousands of onlookers with an unannounced lift-off, he flew the Flying Boat at 70 feet for one mile before making a perfect landing.

Half a century later and still the largest airplane ever built, Evergreen was saddled with the challenge of moving it from southern California to Oregon. First, the aircraft had to be removed through the ceiling of the geodesic dome in which it

was stored. A barge and tug boat carried the parts to Portland. At Portland, the cargo waited for water levels to be low enough that it could pass under the bridges on the Willamette River. The fuselage was pulled the last mile over land by three 475-horsepower prime movers, each with 104 forward gears. All told, the 1055-mile odyssey took 138 days.

Now, the museum is open seven days per week and most holidays. The museum is home to 40 historic airplanes, including a B-17G, P-51D, Ford Trimotor, and a homebuilt BD-5. Because many are flyable, expect some to be unavailable for viewing when you visit.

Entry is $9 for adults, and less for children, seniors, students and veterans. For information, call 503-434-4180 or see their web site at *www.sprucegoose.org* .

- RW

Historic photos: Evergreen Aviation Museum

OR - MEMA

Memaloose

Memaloose Airport

Town	Imnaha **ID** 25U
Coord	N-45-25.66, W-116-41.63
Elev	6708 feet
Runway	18rt - 36rt, 3300 x 120 dirt
Freq	CTAF-122.9
Charts	Great Falls, Seattle sectionals

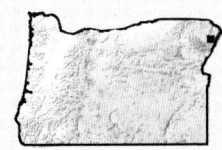

CAUTION: *Airport information not for navigational use. Damp spot at north end in spring, so land long to south. Usable length is 3000 feet. Livestock on runway. Watch density altitude.*

> *Memaloose is a top-of-the-world campsite.*
> *Spectacular views of the Snake River Canyon.*
> *Ideal for portable ham radio operation. $0*

Airport Description

Look for Memaloose just west of the Snake River Canyon on the Oregon/Idaho border. The southern tip of the runway ends at the edge of a cliff that drops into the canyons of the Snake River. A white windsock is located at the southeast corner of the field. White rocks outline the runway at intervals. The markings are difficult to spot from a distance.

A dirt road runs parallel and east of the runway. Do not land on the road – the runway is smoother. Either end can be muddy in spring or after prolonged rains. I watched a German visitor bend his prop in a soft area at the south end. For several seconds, his Cessna 182 could have passed for a 235 HP manure spreader.

Although the runway has rough spots, it has always been solid for my late-summer visits. Unless the wind strongly favors otherwise, land south and depart north.

I was unable to locate an established tiedown area. Bring your own ropes and stakes, and tiedown most anywhere.

Services and Lodging

Phone: USFS, 541-963-7171 (not at airport). No services, fuel, bathrooms, power, nor phone are located at the airfield. Camping is permitted, but no improvements are located at the strip. A number of idyllic campsites are available close to the runway. Bring your own water. A campground with improvements is located two miles northeast at the fire tower.

© 2001 by ALTA Research

Memaloose

Overview

When I walked the edges of the plateau, I could not help feeling that I was walking on top of the world. Except for the nearby fire tower, no peaks rise higher than the strip – you've landed on the top. Walk to the south end of the runway and look down several thousand feet into the Snake River canyon.

I selected a pleasant campsite near the southeast corner of the strip. At this site, I inherited an informal fire-ring and a soft natural bed surrounded by a protective family of fir trees. This location placed me within two minutes of my plane and spectacular views of the canyon. I made frequent visits to the edge to enjoy nature's gift. With a full moon in season, I was treated to awesome, humbling sights. My moonlit hike along the rim leaves lasting mental images that make me smile as I write.

Still smiling, I remember two other moonlit personalities that visited each night. I first encountered these local residents shortly after securing my plane. Two curious horses, never more than 20 feet apart, cautiously approached the plane. Their curiosity was strong, but not quite so strong as their apprehension of me. With uncanny precision, they maintained their distance at 75 feet. When I was safely distanced, they turned to the plane

and gave it a close inspection. We kept to ourselves until later that evening.

The moon rose and I felt inspired to play a melody of tunes on my harmonica. After playing several standards, I heard heavy footfalls and became aware of two large black hulks moving slowly in my direction. Delighted to have an audience, I stood up and played a few favorites – with feeling! After my third, four large eyes were not more than one foot from my face. As they let me stroke their velvet noses, they breathed warmth into my chest. I felt like the Pied Piper.

This special moment was made at the expense of privacy. My new friends became frequent visitors, clomping through camp without invitation, poking noses into whatever, and stepping on gear about the site.

Ham Heaven

This locale is perfect for portable amateur radio operators. A "ham" myself, I enthusiastically slingshot a 40 meter dipole

QSL from Memaloose, OR - 40M, CW, 2 watts

antenna into the tall firs. I easily made contacts over 2000 miles away with a battery-operated, 2-watt, Morse code transceiver.

Fire Tower Hike

This loop takes you along the edge of the plateau to a fire tower and campground located two miles northeast of the strip. You can walk the loop in two hours.

From the southeast corner of the airstrip, walk counter clockwise along the rim of the plateau. You'll see exceptional views on most of the journey to the tower. Take time to enjoy the local color and flora, present even into fall.

The tower is 75 feet high. Visitors may climb the tower at their own risk. The view from the top is 360 degrees of spectacular corrugated topography. Water and toilet facilities are available at the adjoining campground.

As you return, follow the same road about half way back. Use your pilot's sense of direction to bear right and continue on the road that loops you counterclockwise back to the strip.

OR - MINA

Minam Lodge

Minam Lodge Airport

Town	Cove **ID** 7OR0 (7-oh-R-zero)
Coord	N-45-21.04, W-117-38.19
Elev	3589 feet
Runway	18-36 est., 2600'est. x 40' turf
Freq	CTAF-122.9
Charts	Seattle sectional

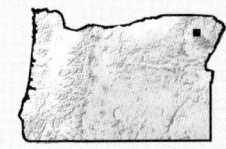

> *CAUTION:* Airport information not for nagivational use. Dangerous mountain terrain. Limited maneuvering space. Nearby trees. Drafts. FAA information may be out of date.

Horse Ranch

36

Minam Lodge

18

Minam Lodge **OR - MINA**

Minam Lodge is a wilderness guest ranch, offering friendly atmosphere to daytime drop-ins or overnight guests. Hunt, fish, ride, or kick back. For experienced mountain pilots only. $0-300
888-454-4415, www.minamlodgeoutfitters.com

Airport Description

Minam Lodge lies in a narrow valley with steep mountain sides on either side of the river. You'll see two strips within a mile of each other, retired Horse Ranch **OR-HORS**, and Minam. Setup your approach for Minam, the north-most strip.

The Horse Ranch is closed, and has been sold to the U.S. Forest Service. Although ranch facilities are no longer available to the public, its runway was still open in 2001. If this runway appears more negotiable than Minam, consider it an alternative.

The Horse Ranch runway, an alternative just south of the Minam

Flatland pilots will feel uncomfortable here, and should get mountain training before visiting either strip. I fly a good deal in the mountains, yet when I first visited, I did a go-around before executing a satisfactory final approach for the Minam runway.

On the morning I arrived, the air was calm. To setup for a rapid descent and quick stop, I approached toward the south with higher-than-normal power at lower-than-normal airspeed. As this increases risk in turbulent air, I recall keeping a cautious eye on the treetops 75 feet below as I approached from the north. Touch down at minimum speed near the approach end, and you'll have plenty of room to coast to a stop. Tiedown at the center-west edge of the strip.

Most pilots land south and depart north. The ranch recommends early morning and late afternoon operations. Visit only if you have a safe margin of power.

Airport Services

Airport phones (not at airport): 888-454-4415 or 541-562-8008. This is a backcountry strip. The nearest road is some 20 miles north. The lodge has a backcountry radio link to their home base, but no conventional phones or fuel are at the site. The lodge is an easy five-minute walk from the tiedown area.

Minam Lodge

Now that the Horse Ranch, **OR-HORS**, has been sold to the Forest Service, Minam Lodge remains the only backcountry site of its kind in the Wallowa Mountains. Operated since 1998 by husband and wife team Otis Waggoner and Teresa Raaf, the lodge is open for meals, lodging, and a hot shower from mid-May until mid-October.

© 2001 by **ALTA** Research

Minam Lodge **OR - MINA**

For no charge, you can land and camp near the river at the south end of the runway. Close by are running water and pit toilet – a generous offering to pilots from the new owners.

But, hey, why do that! For less than it costs you to fly in and out, you can have three square meals and stay in a cozy log cabin with wood stove, hot shower, and flush toilet. The cabins, which can sleep one to three people, were renovated in 2000. An electric generator provides electric power 2-4 hours per day. The cabin option costs around $220 for two, depending on length of stay. They also offer a dorm that sleeps eight people.

Hungry local pilots fly in just about any time during daylight hours. The meals when I visited were above average, as fly-in ranches go. As with many such ranches, meal prices are higher than your typical year round small-town diner: Breakfast and

Time for a good book while the boys are at play

lunch are $10 each. Dinners are $18. Considering that the food and fuel are either flown in or packed in by mules, the prices are reasonable.

Hunters should ask the owners about options for hunting in the area. They can arrange **pack trips**, **hunting trips**, and **fishing trips** from the lodge.

Non-hunters can enjoy the **scenic beauty** of the Wallowas at Minam throughout the warmer months. From the lodge, the owners can provide guided or guide-free **horse rides** that take you up into the mountains. My group especially appreciated being trusted to find our own way in and out of the mountains. The cost per mount for an unescorted ride was $25 for two hours or $40 for four hours.

The clank of horseshoes, a familier sound before dinner

Minam Lodge **OR - MINA**

There are lots of options for stream, river, or lake **fishing**. Fishing along the Minam river is outstanding. During my first visit, I watched a fisherman pull in several large Rainbows from Beaver Lake, just 20 minutes from the Lodge.

For additional information about the Minam Lodge, call the numbers listed under "Airport phone" above, or write to Minam Lodge, PO Box 3384, La Grande, OR 97850. Check out their web site at *www.minamlodgeoutfitters.com* .

Maps are available from the U.S. Forest Service at 541-426-4978. The Fish and Game Department is at 426-3279.

At Wallowa Lake, similar hiking, packing, hunting, and fishing activities are offered by outfitters. See **OR-WALL** for other access to the Wallowa Wilderness.

- RW

Visitors saddle up for an afternoon ride

OR - NEWP

Newport

Newport Airport

Town	Newport **ID** ONP (oh-NP)
Coord	N-44-34.82, W-124-03.48
Elev	160 feet
Runway	16rt - 34, 5398 x 160' asphalt
	2 - 20rt, 3001 x 175' ends low
Freq	CTAF-122.8, Lights-2x5x7x
	FBO-128.95
Charts	Seattle sectional

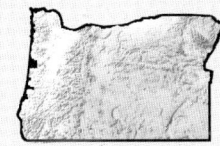

CAUTION: Airport information not for navigational use. Hills E, and rising terrain N. Birds, deer, and helicopters.

> *Newport is a scenic coastal town with ocean fishing, whale watching, and a world-class aquarium. Lively bayfront, and lots to see and do in the area. $100-250*
> *800-262-7844, www.newportchamber.org*

Airport Description

The Newport Airport is a mile east of the Pacific Ocean, 2 miles south of Yaquina Bay, and 3 miles south of Newport. The coast experiences overcast patterns at various times of day and year. Best hours for arrival are 11:00 to 18:00. The airport's ILS, VOR, NDB, and GPS approaches can come in handy.

Services, Transportation, and Lodging

Airport phone: Central OR Coast Air Services, 800-424-3655. FBO frequency: 128.95. Fuel: 100 and Jet. The FBO is a full-service FBO, offering maintenance, flight instruction, and transportation. They have a courtesy car for short runs, and rental cars for sightseers.

Taxi: 541-265-9552. The FBO provides rental cars from Enterprise for $43-100. Make arrangements in advance through Air Services. For lower rates, you can pick up a car from Enterprise in town, 574-1999. Robben Rent-a-Car, 994-2452, picks up from the airport and rents cars for $30-up.

A Newport shuttle trolly carries passengers between north and south sides of the bay, but does not stop at the airport. The Embarcadero Resort Hotel & Restaurant provides shuttle support to destinations on both sides of the bay. It also rents bicycles and sea kayaks for use in the bay.

Newport

OR - NEWP

There are three interesting areas in town, and places on the coast to visit. You will want to visit the historic north bay front, the south bay peninsula, and the Nye Beach area, northwest of the north bay. I recommend renting a car.

Features
B Full breakfast
C Continental
H Historic
J Jacuzzi
P Pets OK
R Restaurant
S Swim pool
T Transportation

Newport has over 80 restaurants and three dozen options for lodging. If you arrive without reservations and are unable to find lodging at the better-known inns, try one of the local competitive mom and pop motels.

Name	Features	Price	Phone
Agate Beach Inn B.W.	JPRST	$70-160	800-547-3310
Beach House B&B	BJPT	$90-130	866-215-6486
Embarcadero	JST	$80-180	800-547-4779
Green Gables B&B	BJT	$90-130	800-615-9065
Hallmark Resort	JPRST	$130-175	888-448-4449
Hawthorn Inn & Suites	BJPST	$70-170	800-527-1133
House Rogue Bed & Beer	HPR	$80-130	541-265-3188
Newport Belle B&B (boat)	BT	$85-155	800-348-1922
Nye Beach Hotel	JRT	$55-160	541-265-3334
Ocean House B&B	BJT	$96-245	800-562-2632
Shilo Inn	PRST	$75-200	541-265-7701
Sylva Beach Hotel	BHR	$68-180	888-795-8422

Newport

Anchored by its slogan, "Preserving the Treasures of the Oregon Coast," Newport has become a favorite destination for visitors in Oregon. The town offers historical sights, seafood treats, ocean adventures, a world-class aquarium, and scenic drives along the coast.

While growth has been stimulated by tourism, Newport's Yaquina Bay still supports a fishing fleet and small seafood processing factories. It also has a wood products industry, a

© 2001 by ALTA Research

microbrewery, and the Oregon State University Marine Science Center. Impressive, for a town of 10,000.

Originally, Newport was home to the Yacona Native Americans. In the mid 1800s, the white man began visiting the beautiful area. Some settled down. The Federal government designated Yaquina Bay as Indian reservation, but reversed their decision after the discovery of local oysters.

Now a historical museum and gift shop, the 1871 **Yaquina Bay Lighthouse** is Newport's oldest building. Because its beacon was found to be blocked for ships arriving from the north, the 1873 **Yakina Head Lighthouse** was rapidly built 3 miles north. The latter, 93 feet high and visible 19 miles at sea, is Oregon's tallest and second oldest active lighthouse.

Nye Beach became a hot tourist destination in the early 1900s. Today, you would not guess that Nye Beach had hosted penny arcades, curio shops, salt water taffy parlors, and such.

Newport **OR - NEWP**

All that and more migrated to the bayfront years ago. Attractions like Undersea Gardens, a wax museum, and Rippley's have arrived to offer their forms of entertainment.

The **bayfront** is a good place to begin exploration. Finding a place to park near the one-mile strip can take patience. You pass interesting stores, **art galleries**, **bars**, **seafood restaurants**, **fishing** and **charter piers**, canneries, and attractions designed to amuse tourists. Artists have used building-sides as their canvas for gigantic murals. Scents of cooking seafood float by.

Why resist? The seafood meals are fresh and affordable – you have your pick of restaurants and bars. Once a local favorite, Mo's restaurant holds the dominant location on the strip. Other establishments, such as the nearby Whale's Tale, serve a satisfying meal. Just up the hill from the bayfront, the Canyon Way Restaurant & Bookstore provides fine food in a captivating environment, at prices somewhat higher than the others.

It is not uncommon to see people catching crabs right from bayfront piers. Within a half hour, I observed several people pull in crabs from a pier near Mo's. The Embarcadero Hotel & Marina allows guests to catch their own crab from their private dock, and then cook the catch on site.

One mile northwest, **Nye Beach** is again becoming the place to stay and eat in Newport. Many of the old buildings have been renovated to provide tasteful lodging and dining with laid-back ambiance. You can get a great bowl of chowder at the Chowder Bowl, or a nice meal with ocean view at the Nye Beach Hotel. April's and Arr Place are two other fine Nye Beach restaurants.

The Oregon Coast Aquarium, OSU Marine Science Center, and **Rogue Brewery** are on the south bay. After several hours of science, might I suggest a visit to the brewery. To get to the bar and grill, you walk through the center of the brewery, up some stairs, and into a rustic loft that has views of the bay. From this perch you can watch fishing boats come and go while washing down a tasty meal with a sampling of Rogue beers. Their bayfront pub across the bay has music and dancing after 10:00pm.

Newport's premier attraction is its **Oregon Coast Aquarium**, 541-867-3474. Opened in 1992, the aquarium soon achieved international fame by restoring the health of Keiko, the killer whale who had seized the hearts of children and adults in the movie *Free Willy*. After his recovery, Keiko was returned to his native waters of Iceland. His large pool has since been converted to include a 200-foot series of transparent tunnels, though which visitors can walk and observe sea life in all directions.

© 2001 by **ALTA** Research

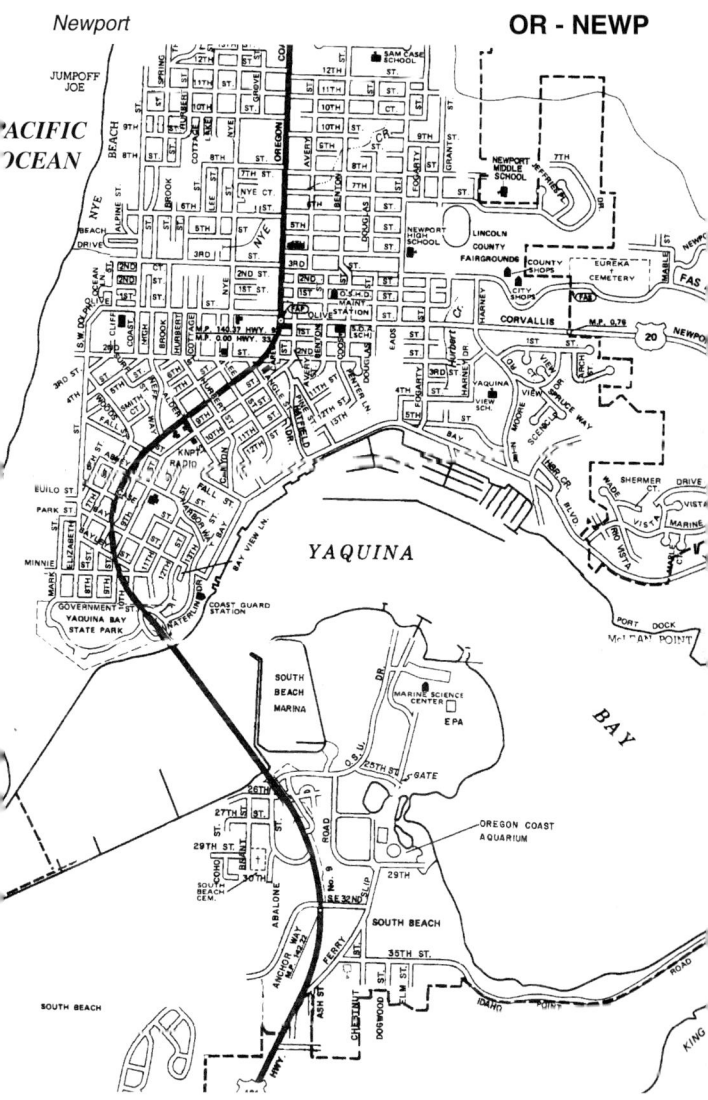

With over 10,000 marine animals representing 500 species, the aquarium replicates Oregon coast habitats above and below water. Outdoor visitors can see seals, sea lions, sea otters, and birds. Underwater portals reveal fabulous color and animation of the life below – sharks, brightly colored fish, tidal pool organisms, and more. One of my favorite finds was the aquarium's etherial display of delicate lively jellyfish from all over the world. The aquarium is a class act. Admission is $9.50, and less for children.

The nearby **Oregon State University Mark O. Hatfield Science Center** provides another slant on marine science, 541-867-0100. The public portion focuses on the research work of the center's 300 marine scientists. Visitors learn through interactive, multimedia displays. The center includes a touch tank and live octopus. The entry is a $5 donation.

To the north, Yaquina Head is home to the **Yaquina Head Lighthouse and the Interpretive Center**, 541-574-3100. To

Newport **OR - NEWP**

gain access to this BLM facility, citizens now pay an additional Federal tax of $5 per car. The area around the center provides good places to photograph the lighthouse, glimpse gray whales, explore tide pools, and observe sea birds and harbor seals. Bring binoculars and long camera lens.

History buffs should check out the museums in town. The **Yaquina Bay Lighthouse** of 1871 has been restored into a museum and gift shop. Reputed to be haunted, it has a legacy of ghosts and mystery dating to 1899. The **Log Cabin Museum & Burrows House Museum** focuses on the history of pioneering, the local Siletz tribe, and logging and fishing industries of Lincoln County, 541-265-7509. No mandatory fee.

Newport has a choice of **bay and deep ocean activities**. Marine Discovery Tours, for example, gives slow or high speed **boat tours**, where you may see seals, sea lions, porpoises, whales, and seabirds. Prices are affordable. Charter services provide **fishing and crabbing trips**.

Bayfront Charters	*fishing*	800-828-8777
Marine Discovery Tours	*sea life cruise*	800-903-BOAT
Newport Marina & Charters	*fish, rent*	877-867-4470
Newport Tradewinds	*fishing*	800-676-7819
Sawyer's Landing	*crabbing, rentals*	541-265-3907
Sea Gull Charters	*crabbing, fishing*	800-865-7441

Events

Call the Chamber for dates and details.

Feb,mid	Seafood & Wine Festival	800-262-7844
May,begin	Loyalty Day and Sea Fair	541-867-3798
Sept,begin	Rhythm by Bay Music Fest	541-867-3798
June,mid	Gem and Mineral Show	541-867-6903
Oct,begin	Rock Rhythm & Blues	503-241-3800
Dec,end	Whale Watching Week	800-262-7844

Oregon Coast Aquarium

© 2001 by ALTA Research

OR - OWYH

Owyhee Reservoir

Owyhee Reservoir Airport

Town	Owyhee Reservoir **ID** 28U
Coord	N-43-25.49, W-117-20.73
Elev	2680 feet
Runway	12 - 30, 1840 x 30' dirt
Freq	CTAF-122.9
Charts	Klamath Falls sectional

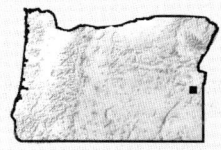

> ***CAUTION:*** *Airport information not for navigational use. Expect sink over water on approach. Runway not perfectly flat, expect hills and valleys. Occasional rocks on runway.*

> *Lake Owyhee, a quiet place, undiscovered by time and money – a shrine for those who mourn erosion of freedom, accountability, and integrity. A modest dirt strip at the edge of a lake, not accessible by land vehicles. Enjoy the simplicity of a rent-free shack, stocked with utensils, food, and libations from previous visitors.* $0

Airport

If you can locate Lake Owyhee, a long north-south reservoir near Idaho, you will have little difficulty finding the runway. Look for a dirt strip on a peninsula on the western shore, with a cabin 1000 feet to the east. The approach is straightforward. Be prepared for possible sink over water and expect manageable hills and valleys in the runway. The dirt/gravel runway is normally dusty with occasional larger rocks. Expect mud after a downpour. The runway was more than adequate for my C-172.

I counted at least eight tiedowns, one of which is conveniently close to the cabin.

Services and Lodging

Phone: Oregon Aeronautics, 503-378-4880 (not at airport). The airport is not attended and there are no modern conveniences like phone, potable water, or power. Bring your own water or filter. If you opt to filter lake water, I recommend a two-stage filter because of a high particulate level in the lake. Also, bring a shovel for creating a mini-latrine.

The most unusual service at Owyhee is the free tin cabin, which is perpetually stocked with provisions by visitors.

Owyhee Reservoir **OR - OWYH**

Owyhee Lake

Owyhee Lake is a low-use desert-like recreation area near the eastern border of Oregon. The beauty of the fly-in site is subjective, and can be judged only by the eye of the beholder. Some see paradise; others see a run-down shack at the edge of a dusty airstrip.

You be the judge, but let me share my perspective. My eyes saw an escape from a world of regulation, rules, fears, etc., and at this site, a return to old values that struck a chord in my heart.

As I write this chapter, I sit in a lovely beach house at the southwestern tip of Puerto Rico. The people are friendly, yet you need to keep everything locked up tight. The beach is beautiful and uncrowded; the weather is perfect. So I ask myself, why do I feel this affinity for the shack on the lake?

Owyhee Hilton

Owyhee is free. True, but many remote western strips are free. Owyhee reminds me that life can be sweet without material excesses, except, of course, for my trusty Romeo, N3790R. Owyhee brings out the best in her visitors. As documented in the cabin's log, visitors forget their worries, laugh together, fraternize with the local Owyhee Nymph, tidy up when time to leave, and leave gifts for the next visitors.

Cave campsites, petroglyphs, and arrowheads in the valley indicate that humans have visited the area for 12,000 years. The name "Owyhee" commemorates two Hawaiians hired by early beaver trappers who were killed there by Indians in 1819. "Owyhee" is a 19th-century spelling for Hawaii.

In foul weather after a forced landing, the little tin cabin would be a welcomed refuge. In better weather, the accommodations are clearly funky and in need of cleaning and repairs. Rather than sleep inside, I preferred to sleep on the wooden porch where I enjoyed scenery, stars, and a small campfire. Note that wood may be scarce; fly some in, if you can.

After sunrise, the shaded porch and trees provide a welcomed retreat from the sun. Few experiences can top moments like sitting on the back porch, the feeling of a warm breeze tickling your armpits as you lift a jug of wine. Add cheese and crackers and you have instant paradise.

The porch overlooks an eastern view of lake with a backdrop of rugged bluffs and outcroppings – a tempting destination for a day's adventure. The cabin holds a rubber raft to help launch the expedition. The lake is shared by a small number of other vacationers, who may wave as they zoom by.

Fish runs on the Owyhee River died when the Owyhee Dam was built in the 1930s. The reservoir now teems with black crappies, and the river supports small populations of catfish,

Owyhee Reservoir **OR - OWYH**

suckers, and carp – not exactly gourmet fare. Besides catching a few, the area offers **exploring**, **mountain biking**, **plinking**, **hunting**, or searching for **arrowheads**. Watch where you walk; I can vouch for the presence of rattlesnakes.

The cabin is stocked with a variety of canned foods and an unusual collection of cooking utensils. In groping through the cache, I discovered a half-full bottle of Cuervo Gold that, according to the log, had been left several days earlier by a group passing through from Mexico.

My most recent visit, July 4, 1996, showed not much had changed. The water level was high and lots of driftwood had been left on the shore. The cabin had its usual cache of food – this time with Reposado Tequila Herradura. Someone had filled the cabin's cracks with insulating foam and patched the bullet holes with duct tape.

Take your siesta here, and the Owyhee Nymph will surely visit

The cabin logs are a precious asset. Take some time before siesta to browse the log to see what previous visitors have said about the Owyhee Nymph. Slip into siesta, and soon she will visit you. Then you, too, can make an entry like the following:

"The Hood River Bunch came. We tried and were conquered by the Owyhee Nymph and her wicked ways. Twelve men came, twelve men left, thankful that wives and decent citizens weren't present to witness the travesty that took place on this annual Memorial Day trip.

"So we leave this wondrous place after being drenched by a two-day downpour, the likes of which have never been seen here before or will be seen here since the dinosaurs have roamed these hills.

"Perhaps the Big Guy that dwells above where even our turbocharged little craft can't go got a message from his friend, the Owyhee Nymph, and decided to drown these amorous men with a deluge that nearly drowned man and beast, alike.

"We find it fitting that such tale of truth be the last one in this log book. It would also be our hope that future travelers might learn from this sordid tale and conduct themselves with more dignity and restraint than did we, or they will surely get wet. So long, Hood River Travelers, 5/25/90."

Breakfast in Rome

When it comes time to leave Owyhee and Nymph, why not plan an early-morning flight to Rome for a hearty breakfast? Fly about 40 miles south and look for the large dirt strip next to the river. The strip is marked as "Restricted" on the sectional. It is an easy landing, but as most restricted strips, it is not insured.

You can park within several hundred feet of the RV park coffee shop. The owners serve an honest breakfast at a reasonable price.

- RW

June 19

What a splendid p[lace to]
set up camp. As we [ ]
away the hot afternoo[n]
saw a movement [ ]
opposite shoreline [ ]
tanned outline a[gainst]
the white limestone [ ]
attention. As I g[ ]
panes of bin[oculars]
glass I [ ]
this a[ ]
I be[ ]
is anot[her]
confirm[ed]
of the el[usive]
wood Ny[mph]
I must fi[nd]
wood to bu[ild]
as I plan to [ ]
beating this [ ]
do if you fi[nd]
is hard to come by
blame the drough[t]
lack of trees, ju[st build]
a raft and [ ]
can fault the [ ]
and certainly m[ ]
Phil Quayle 6-1[ ]

WANTED
MISSING PERSON

Information leading to the where abouts of the OWYHE NYMPH

Last seen clad in rabbit skins traveling with a half crazed pilot over the east hills

$1000 REWARD

OR - PINE

Pinehurst Box R Ranch

Pinehurst State Airport

Town	Pinehurst **ID** 24S
Coord	N-42-06.49, W-122-23.07
Elev	3650 feet, SW end high, 4%
Runway	4-22, 2850x30, NE end paved
Freq	CTAF-122.9
Charts	Klamath Falls sectional

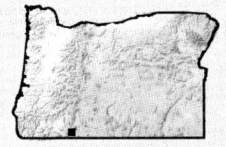

CAUTION: Airport information not for navigational use. Call 503-378-4880 before landing. SW end is rough and somewhat higher than NE end. Trees, livestock. Watch density altitude.

> *Pinehurst has a guest ranch and B&B inn set in the Siskiyou Mountains. The ranch offers summer horseback riding, daily hay rides, hiking, and fishing. $80-175*
> *Ranch: 541-482-1873, www.boxrranch.com*
> *B&B: 541-488-1002*

Airport Description

Pinehurst sits in the Siskiyou Mountains, 14 miles east of the Ashland airport. The windsock is easy to spot at the center of the airport. Unless winds aggressively favor otherwise, land on Runway 22 and depart on Runway 4. The strip has no lights and is closed in winter due to snow.

Box-R Ranch pasture

Pinehurst - Box R Ranch **OR - PINE**

Oregon Aeronautics requests that you call them at 503-378-4880 before landing. You can call the Box-R Ranch for current conditions. The airport is unattended and has no facilities. There is no phone, and cell phone coverage is spotty.

Several tiedowns are located at mid-field. Most of the land around the airport is BLM (Bureau of Land Management) land. Camping is permitted on BLM land, but most visitors prefer the comforts of the ranch or B&B.

Transportation

The Pinehurst Inn is within 15-minute walking distance, and the Box-R Ranch is within 20 minutes. Both will pickup from the airport if notified in advance. When you arrive, buzz the appropriate facility, or try calling with a cell phone.

Green Springs Box-R Ranch

The Box-R Ranch, 541-482-1873, is a **working cattle ranch** of 1,000 acres that hosts families and groups year-round. Located in a high mountain meadow surrounded by forests of pine and fir, the ranch's crisp air and no-alcohol policy will breathe new life into family relationships.

Guest houses are nestled among the pines with views of the lake and meadow. Each house has two bedrooms, a modern kitchen, washer, and dryer. Their spacious living rooms all have a rock fireplace. Rates are $125 for the two-bedroom units and $175 for a larger version.

Overall, the facilities are ideal for fly-ins, family reunions, and church retreats. Bring your own groceries or dine in style at the Pinehurst Inn.

Box R wagon ride

Pinehurst - Box R Ranch **OR - PINE**

The ranch was brought to its high standard by owner Don Rowlett, a genius from the business world who had a dream and made it happen. He has succeeded in creating an idyllic, wholesome environment that will soothe the souls of all but the most hardened city folk.

The Rowlett family recently opened a beautiful **museum** at the ranch, and grounds are an **open-air museum** in their own right. As I rounded each bend on the trails at the ranch, I spotted antique wagons and Western artifacts seemingly at home in their natural environment.

By reservation, you can arrange for a **wagon ride** that covers the grounds. During summer, **day rides** from nearby stables follow trails that venture into the wilderness. Other activities include **swimming**, **fishing**, **hiking**, **mountain biking**, **volleyball**, and a variety of **winter activities**.

Box-R Museum

Pinehurst Inn B&B

The Pinehurst Inn is a charming bed and breakfast inn with excellent meals. According to local lore, the inn is a restored 1920's roadhouse that once served duty as a bordello. Now, respectable guests enjoy the historic rooms that sit over a gourmet restaurant below. On weekend mornings, local pilots join the respectable folks for a tasty breakfast.

Prices for two at the inn range from $79 to $99, including dinner and a deluxe continental breakfast. Walk-in visitors are served dinner for $12 each. The inn has switched to new management since my last visit. If the tradition of quality has been maintained, as I expect it has, you will enjoy your stay.

The Pinehurst Inn is a nonsmoking environment. Your credit card will hold a reservation, but the terms have various contingencies. Call 541-488-1002 for details and reservations.

The nearest local attraction is the Box-R Ranch. Local activities include **hiking**, **fishing**, **white river rafting**, and **pack trips**, and more.

Ashland, home of the **Shakespeare Festival**, is a half-hour drive. See chapter **OR-ASHL**. Less than an hour's drive is historic Jacksonville, an 1850s **gold rush town** and home of the **county museum** and **Britt Music festival**. Ask the innkeepers to help with arrangements for any of the above attractions.

Contact the inn in advance to alert them of your arrival. The inn is not far from the northeast end of the runway. On prior arrangement, give it a good buzz before landing. Or, simply walk. From the airport, head north to the highway, turn right, and follow the highway east to the inn.

- RW

© 2001 by ALTA Research

OR - SANT

Santiam Junction

Santiam Junction Airport

Town	Santiam Junction **ID** 8S3
Coord	N-44-26.07, W-121-56.54
Elev	3770 feet, E end high, 1.5%
Runway	6 - 24, 2800x150' dirt/rock
Freq	CTAF-122.9
Chart	Klamath Falls sectional

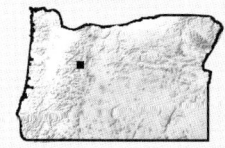

CAUTION: Airport information not for navigational use. Call 503-378-4880 before use. No go-around for Runway 6. Do not depart 6. Rising terrain nearby.

> *Santiam Junction is an emergency strip in the middle of the Cascade Mountains. No place to spend your money – a nice quiet place to hike. $0*

Airport Description

The strip is located just south of the highway maintenance facility at the intersection of Highways 20 and 22. The runway is wide and generally maintained rut-free. High terrain surrounds the airport (particularly at the east end), so use caution. The east end slopes up. Land east and depart west.

Apparently the strip is used primarily for emergencies. A group of children surrounded the plane immediately after I arrived. They claim to have seen only two other aircraft land at Santiam all year – two helicopters. The airport is closed in winter.

Airport Services, Transportation, and Lodging

Oregon Aeronautics suggests that you call them for runway conditions at 503-378-4880 before landing. A pay phone is located on the highway about one-quarter mile west of the junction, not far from the airport. No fuel is available. Bring your own tiedown hardware. No food services are close to the airport, but water is available.

Santiam Junction **OR - SANT**

Overview

The scenic mountain terrain surrounding Santiam Junction offers variety, both in type of terrain to cover, and in ease or difficulty of travel. Expect to encounter some cross-country travel. Trails are not as clearly identifiable as they are in officially designated wilderness areas. The lack of maintained trails offers a challenge, as well as the benefit of rarely meeting other hikers.

A topo map and compass are musts. I recommend an accurate compass with mirror. The two-degree accuracy of a mirrored compass is valuable for triangulation fixes. Before landing, identify and note the profile of prominent mountains such as Hoodo Butte, Hayrick Butte, Nash Crater, Mount Washington, and Three Fingered Jack. These landmarks will prove useful for triangulation.

Hiking to the southeast near Big Lake is relatively easy. The Pacific Crest Trail (labeled *Skyline Trail* on the topo) runs north-south four miles east of the airport. The highlands, several miles northeast of the airport, are inviting and relatively untouched by man. The Maxwell Butte Loop (explained below) provides penetration into this area. You can extend the loop into the Duffy Lake area for a picturesque several-day hike.

The terrain in all directions lends itself to the creation of a custom hiking adventure. Armed with good maps and gear, you are ready to go. Consider the following Maxwell Butte Loop if you prefer to avoid seeing other hikers.

© 2001 by ALTA Research

Maxwell Butte Loop

This loop follows unmarked trails, and is recommended for hikers who are proficient at wilderness orienteering. A compass and the *Three Fingered Jack* topo map are necessary. Count on losing the trail, and gaining the opportunity to use your orienteering skills. Besides the satisfaction of depending upon personal skills, you will be spared the sight of other hikers in this area.

From the landing strip, proceed to the pay telephone that is located on the highway, one half-mile west of the junction of Highways 20 and 22. From the phone, cross the highway and walk approximately 1000 feet due north until you intersect the old logging road. Follow it northeast for about thirty minutes until you reach the volcanic rock-encrusted saddle between two knolls.

Soon the logging road joins an old railroad grade. Follow what's left of the railroad bed for ten to fifteen minutes after the saddle, then turn left on the well-defined trail that heads up the mountain. The next fifteen minutes are the steepest climb on this loop.

Upon reaching the ridge, continue to follow it in a northeast direction for about one hour. For the duration of this hour, terrain is a mixture of trees, rock, and grassy meadow. From time to time, mountain peaks show themselves between the trees.

Using triangulation, carefully determine your position from other mountain peaks. In order to take accurate compass readings, be sure to separate your compass from backpacks and other steel by a dozen or more feet. Your position at this time should

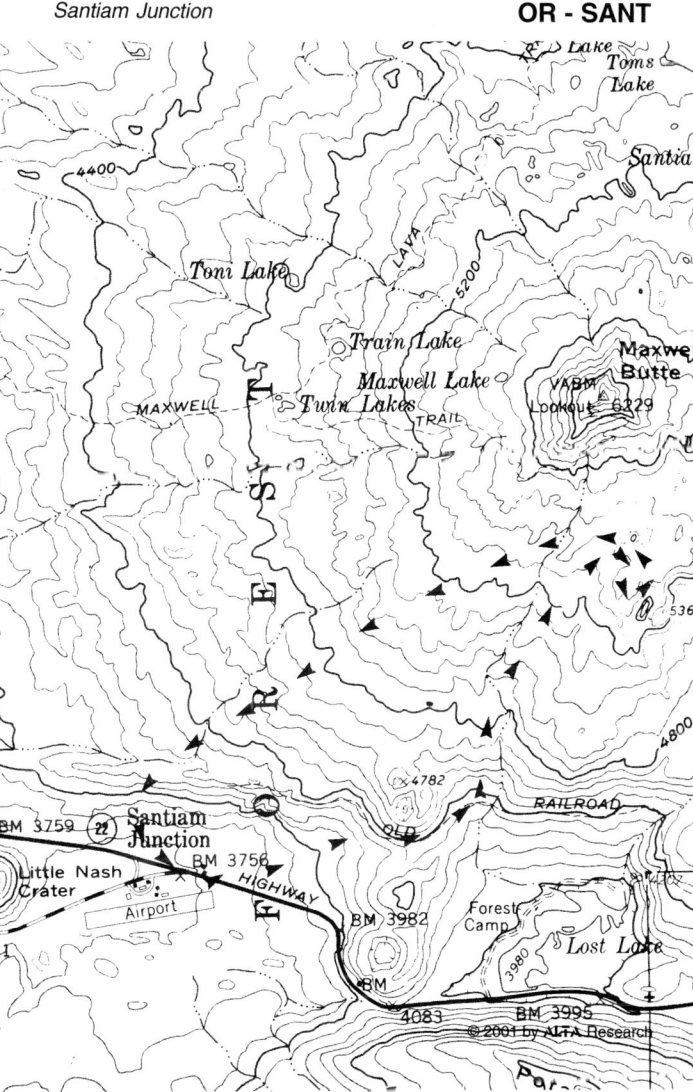

be approximately three-quarters of a mile south of Maxwell Butte. Depending upon time available and personal taste, select an interesting destination from the topo map. Comfortable, unblemished campsites can be found at a number of small lakes in the area.

When returning to the airport, don't be surprised if you have difficulty rediscovering the trail. Simply follow the compass needle southwest toward Santiam Junction. Cross-country travel is generally pleasant, with the exception of the steep, wooded terrain a half-mile north of Santiam Junction.

- RW

OR - SEAS

Seaside

Seaside Municipal Airport

Town	Seaside	**ID** 56S
Coord	N-46-00.99, W-123-54.33	
Elev	6 feet	
Runway	16 - 34, 2357 x 50' asphalt	
Freq	CTAF-122.9	
Charts	Seattle sectional	

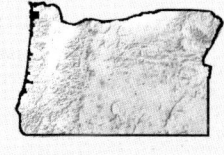

CAUTION: Airport information not for navigational use. Phone line at each end. Elk. Wind shear and variable winds. Sink on approach.

> *Seaside is an oceanside amusement town with a touch of quality above the norm. A fun family stop for day or overnight visit. Travel light, most everything is within walking distance. $100-175*
> *Info: 800-444-6740, www.seasideor.com*

Airport Description

Seaside Airport is one mile northeast of town, and easy to spot once you have located Seaside. When landing, take precautions for wind shear and sink. The strip has three windsocks. When I landed, all three pointed in different directions. The runway is lighted with low-level lights, but take care of neighboring hills and coastal weather when arriving after dark.

Services, Transportation, and Lodging

Airport phone: 503-738-5511, 738-5112 (neither at airport). Fuel: None. Facilities are limited to Porta-Potty and pay phone.

Town center, at Highway 101 and Broadway, is a 30-minute walk from the airport. The central beach area is less than ten minutes farther. Seaside Arrow Cab taxi service advertises at the airport, 503-738-5252 or 738-4005. Rental cars are no longer available.

Because of its Convention Center and recreational draw, this small town has forty motels, eleven B&Bs, and fifty vacation home rentals. If a large convention is under way, lodging can be a problem. Call the Seaside Visitor's Bureau at 800-444-6740 for help. A short list of acceptable lodging follows:

Seaside **OR - SEAS**

Best Western Ocean View	**JPRS**	$96-275	800-234-8439
			503-738-3334
Custer House B&B	**BHT**	$70-80	800-738-7852
			503-738-7825
Ebb-Tide Motel	**JS**	$50-170	800-468-6232
			503-738-8371
Four Winds Motel	**T**	$86-112	800-818-9524
			503-738-9524
Gilbert Inn B&B	**BHT**	$95-135	503-738-9770
Microtel Inn & Suites	**C**	$59-90	888-771-7171
			503-738-8971
Sand and Sea (condos)	**S**	$86-240	800-628-2371
			503-738-8441
Seaside Hostel	**T**	$32-62	503-738-7911
Seasider Motel	**PT**	$37-64	888-070-6544
Shilo Oceanfront	**T**	$85-192	800-222-2244
			503-738-9571

Features
- **B** Full breakfast
- **C** Continental
- **H** Historic
- **J** Jacuzzi
- **P** Pets OK
- **R** Restaurant
- **S** Swim pool
- **T** Transportation

Except for the Seasider, the Hostel, Microtel, and B&Bs, most of the above are on the oceanfront. The Seasider is included here not for quality, but for its near oceanfront proximity and low price. The upscale Shilo has the best location, virtually in the middle of the action. The larger and older Sand and Sea is south. The Best Western is north, with restaurant and kitchenettes. Also north is the Four Winds, a newer facility with appealing architecture that struck me as a cross between Victorian, New England, and Colonial. The Ebb-Tide is a little above the oceanfront norm. The more tacky oceanfront motels are not listed here.

The B&Bs listed above were all constructed at about 1900. Seaside's dozen B&Bs are sprinkled throughout the town. For the lone economy-minded pilot, the Seaside Hostel is an option.

© 2001 by ALTA Research

Dormitory beds cost $15 for members and $18 for nonmembers. Private rooms for two are a little over double single-person prices. Seaside Vacation Rentals can provide a house or condo for large parties, 800-738-776.

Seaside

The first white men to visit were members of the Lewis and Clark expedition in 1805. Seaside is officially the end of the Lewis and Clark Trail. Three of the party spent part of the winter at Seaside making salt from sea water. The first significant building in Seaside was Summer House, an inn built in the mid-1880s by the widow of a Hudson's Bay Company river pilot.

In 1920, Seaside residents enhanced their town's charm by adding a 1.5 mile concrete **boardwalk**. They continue to make improvements for weekenders from Portland and conventioneers. As you stroll the boardwalk and streets, you can enjoy a taste of old and new in this tidy seaside town.

To get to Seaside from the airport, turn left on Highway 101 and head south. After walking five minutes, cross the bridge over Neawanna Creek. After the bridge, you have a choice. You can leave 101 and head toward the beach, or continue on 101 to the Visitor's Center at Broadway. If you choose the latter, continue another 20 minutes to town. At Broadway, turn right, walk past shops and restaurants, cross the river, and you're at the beach.

If you opt for the oceanside route from the southwestern bank of the Neawanna river (near the airport), bear right onto the road that heads toward the ocean. It wraps around and runs parallel to the ocean. When you reach 12th street, turn right, cross the river, and soon you will enter the north end of the boardwalk. Follow the boardwalk south to Broadway, the liveliest part of town. On the way, you pass motels and the Seaside Aquarium.

© 2001 by ALTA Research

Seaside **OR - SEAS**

The boardwalk is pleasant – old style lights for romantic night-time strolls with grass covered micro-dunes between boardwalk and beach. The natural buffer made me feel close to the ocean without beachside chaos in my face. The beach scene is unique. Three-wheeled **beach bicycles** are the toy of choice. You rent them from several locations on the beach. Otherwise, Seaside's beach is like any other popular beach, except the water is too cold for swimming and there's no trash strewn from end to end.

The Broadway area is the **hotspot** for **cruising**, **arcades**, **rides**, **eats**, **shops**, and **mini-malls**. This part of town has a refurbished look, with cute facades, narrow streets, and sidewalk landscaping of flowers and trees. When I visited, cherry blossoms were blooming and the streets were covered by smiling Portlanders who drove a couple hours for sand and sun. Most of the shops cater to tourists: **antiques**, **art**, **kites**, **beach clothes**, **salt water taffy**, lots of food and drink, arcades, rides, etc.

After enjoying a **kite shop** and some strange snacks, I spotted the **Bumper Car** and **Tilt-A-Whirl** facility. The bumper cars are

the same as the ones from my childhood in New Jersey. I hopped into the fray for a ride. For $1.50, I enjoyed dozens of collisions without having to answer to the FAA. Seaside is like the east coast oceanside towns, but with more class and a lot less soul.

Seaside's **Aquarium** on the boardwalk is one of the oldest on the coast, 503-738-6211. They have a dozen trick seals that you can feed yourself. You buy the fish. A wide range of colorful sea creatures are on display. I felt the $5/adult price was high for what you get. Still, it is less expensive than flying and such experiences are a thrill for children. Children can enter for $3.

Those who enjoy history should visit the **Seaside Museum**, 503-738-7065. It costs $2 for adults and children are free. The museum tells the story of Seaside from prehistoric to present times. It is open from 10:00 to 4:00, daily. The **Lewis and Clark Salt Works**, 861-2471, has a replica of the salt cairn used to extract salt at the end of the Lewis and Clark Expedition.

If you stay more than a day, there's plenty to do: **golfing, fishing, crabbing, biking, hiking, horseback riding, tennis,**

and **whale-watching**. The area has four golf courses. Tennis courts can be found at Broadway School and Seaside High. Rent a horse from Westlake Stables, 503-738-6258. If you have a car, drive north to Gearhart, and up the peninsula to Warrenton and **Fort Stevens State Park**. You can drive seven miles south from Seaside to **Cannon Beach** for more beaches and shops.

Events

With its spacious convention facilities and ample lodging, Seaside hosts festivals and conventions year round. A sampling follows. Seaside Chamber: 800-444-6740 or 503-738-6391.

Feb, begin	Coffee & Chocolate Lover's Festival	*yum, yum*
Feb, end	Seaside...and all that Jazz	*jazz all weekend*
June, call	Seaside...and 'awl that Country	*country music*
June, call	SeaPac Amateur Radio	*large Oregon hamfest*
June, mid	Golf Open	*golf tournament*
July, mid	Miss Oregon Pageant	*classic beauty pageant*
July-Aug	Lewis & Clark Historical Drama	*reenactment*
Aug, mid	Budwiser-Seaside Beach Volleyball Tournament	
Sept, late	Seaside Sand Castle Sculpture...	*festivities*
Oct, mid	Oregon Dixieland Jubilee	*many bands, dancing*
Oct, mid	Wine and Food Festival	*local foods and wine*

© 2001 by ALTA Research

OR - SIST

Sisters

Sisters Airport

Town	Sisters **ID** 6K5
Coord	N-44-18.27, W-121-32.35
Elev	3168 feet, S end high
Runway	2 - 20, 3550 x 30, asphalt
Freq	CTAF-122.9
Charts	Klamath Falls sectional

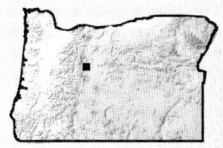

CAUTION: *Airport information not for navigational use. Trees near runway. Strong crosswind common in afternoons. Daylight operations only.*

> *Sisters, where the Cascade Mountains transition to high desert country. The picturesque 1880s style town is a 10-minute walk from the strip. $5-180 CofC: 541-549-0251, www.sisterschamber.com*

Airport

Look for the airport at the northeast edge of Sisters. Be alert for strong crosswinds and possibly wind shear, especially in the afternoon.

Services, Transportation, and Lodging

Airport Phone: 541-549-6011. There are no services or fuel at the airport. The airport is privately owned, but open for public use. A landing fee of $5 is requested, which partially pays the taxes. The runway is frequently plowed in winter.

Transient aircraft parking is available at the southwest end of the runway. The northeast end has a taxiway leading into a 12-lot airpark subdivision called Eagle Air Estates. For information about Eagle Air Estates, call Vern Goodsell at 541-549-9169.

Camping is allowed at the airport. Amenities include shade-trees, tables, and an outhouse. Although the camping facilities are Spartan, they are a godsend for those who fly in without reservations during a town festival.

Transportation is a major limitation at Sisters. If you plan to enjoy the countryside or backpack in the mountains, land at Redmond and rent a car, or ask Enterprise Rent-a-Car to deliver a car to airport or inn. Mountain bikes can be rented in Sisters from Eurosports for $15-25, 541-549-2471. Fortunately, the

Sisters

OR - SIST

town and B&Bs are within walking distance, and one even has bikes for guests.

The lodging closest to the airport is Conklin's Guest House. The B&B is a historic building that has recently been expanded and remodeled into a homey facility of excellent quality. Conklin's has a swimming pool, trout ponds, and massage therapist by appointment. Pets are allowed, but outside only.

Features
- B Full breakfast
- C Continental
- H Historic
- J Jacuzzi
- P Pets OK
- R Restaurant
- S Swim pool
- T Transportation

Best Western Ponderosa	CJSP	$80-100	541-549-1234
Blue Spruce B&B	BJT	$135	541-549-9644
Comfort Inn at Sisters	CJSPT	$75-103	541-549-7829
Conklin's Guest House	BPST	$96-161	800-549-4262
Lazy Rockin' B'n C Ranch	BT	$106-130	888-480-0178
Sisters Historic Mot. Lodge	BPT	$64-107	541-549-2551
Sisters Motor Lodge	BP	$64-107	541-549-2551

Also close to the airport, the Blue Spruce B&B is in Sisters proper. Inns that are less conveniently located at the west end of

Conklin's Guest House

Sisters include: Best Western Ponderosa Lodge, Comfort Inn at Sisters, and Sisters Motor Lodge.

The following are 8 miles or more west of town: Black Butte Ranch, 800-452-7455; and Suttle Lake Resort and Marina, 541-595-6662. At Camp Sherman: Cold Springs Resort, 541-595-6445; Metolius River Lodges, 800-595-6290; and Metolius River Resort, 800-81-TROUT.

The southeast end of Sisters is a half-mile walk from the airport. To get to town, walk south on Camp Polk Road past Conklin's Guest House, and turn right at Highway 20.

Sisters

Stroll down Cascade Street (the main street), and the Hotel Sisters Bar and Saloon is guaranteed to catch your eye. This landmark, the oldest building in town, has not been a hotel since the mid 1980s, but it houses one of Sisters' better restaurants. The back saloon has an elegant mahogany bar that was salvaged from the Palace Bar in Bend, years ago. The wall to the right of the bar has a life-sized mural of high-kick dancing girls, and the remaining walls are decorated with animal trophies and historic artifacts.

© 2001 by ALTA Research

Sisters

OR - SIST

Early settlers predate the 1912 Hotel Sisters by some 50 years. Sisters traces her roots to the settlement of Camp Polk, the remnants of which lie two miles northeast of the airport. Camp Polk was founded by a military expedition in 1865.

The present site of Sisters began at the intersection of two wagon roads that crossed the Cascade Mountains. The settlement officially became an entity in the 1880s when the post office moved from Camp Polk to Sisters. Sheep and cattle ranchers moved their stock through town between summer and winter pastures, until the pastures were closed by the Forest Service.

By 1900, Sisters had become a bustling little community. Timber mills were established to supply ties and bridge members for the railroad between Madras and Bend. The lumber industry grew, becoming the mainstay of Sisters for several decades. The town became known for festive fairs and celebrations until further changes by the Forest Service shut down the mills. The town endured two disastrous fires in the 1930s, and continued to decline until 1970.

Sisters snow storm

The town's luck took a turn for the better when, in 1970, home-sites for the upscale Black Butte Ranch went on the market. Local merchants adopted the 1880s architectural theme, and the town began to change. Face-lifts on the existing buildings were subsidized by Black Butte Ranch promoters.

Today, most every building meets the architectural requirements. The town's western motif stirs a feeling of excitement in her visitors, yet is sufficiently modern to make most city slickers feel at home. Beneath its skin, however, the town has an interesting handicap – no sewer system. For visitors, this means that finding a toilet may require extra initiative. During town events, Sisters beats her handicap by importing enough outhouses to handle the load.

As you stroll the streets, you'll find a variety of **shops to browse**: gifts, antiques, western art, Native American art, fine jewelry, books, recreation supplies, clock shop, a bakery, and more. Fishermen will want to visit the Fly Fishermen's Place for local flies and information, 541-549-3474.

I had fun poking around a used book store on the main street, and scored two interesting finds: <u>Wager with the Wind</u>, a classic biography about Alaskan bush pilot Don Sheldon, signed by Don, himself. And, <u>Listen! The Wind</u>, a 1938 first edition written by Anne Morrow Lindbergh about her flight with Charles Lindbergh in a float plane from Africa to South America in 1933.

Snacks are always within a short walk. Lucky for me, there was no shortage of ice cream shops. For a conventional meal, consider **dining** at the Hotel Sisters, Coyote Creek Café at the west edge of town, or the Gallery Restaurant on Main Street.

The Sisters area offers wonderful recreational opportunities. It was voted as one of the best eight vacation spots in the West by Sunset Magazine in 1993. The Black Butte ranch has a nice **golf course**, and there's lots to do in the Cascade Mountains.

© 2001 by ALTA Research

Unfortunately, the lack of rental or courtesy cars makes these resources difficult to access by pilots.

With transportation, you have more options. The Three Sisters Mountains area is one of my favorite haunts for **backpacking** and modest **mountain climbing**. For information, call the Sisters Ranger District at 541-549-7700. McKenzie River Adventures does **whitewater rafting,** 800-832-5858. The Lazy Z Stables, offers scenic **horseback rides** that cost $20-50. Try these numbers: 541-549-6869, 410-4190, 595-2061, 923-2072.

The **Black Butte Ranch**, 800-452-7455, is located 8 miles west of Sisters on Highway 20. The resort has two 18-hole **golf courses**, 20 **tennis courts**, 4 **swimming pools**, dining facilities, and a wide variety of rental options. In my book, they fall short of world class by not providing transportation from the airport.

Horseback riders enjoy back-country trails, west of Sisters

Events

The first big event begins in early June with the **Sisters Rodeo**, 800-827-7522. This half-century tradition takes place in a full-scale arena 4 miles southeast of town. It is followed on the second Saturday in July by the **Sisters Outdoor Quilt Show**, 541-549-6061, the largest quilt show in the U.S. During this event, the town grows to 25 times its normal size as visitors arrive to view over 600 quilts that have been hung all around town.

In early to mid-August, the **Celtic Festival** is held at Eagle Crest Resort, 541-382-1291. The **High Mountain Dixieland Jazz Festival**, 800-549-1332 happens in mid-September. Ten bands rotate through four venues at the festival. The month's end brings music and storytelling at the **Sisters Folk Festival**, 541-549-4979, *info@sistersfolkfestival.com* .

Nonrecurring events occur throughout the summer. In early October, for example, the **Black Butte Sheepdog Classic** has taken place at Black Butte Ranch, 541-549-7078. Others events include **music concerts**, **classic car shows**, **antique shows**, **craft fairs**, and **rock and gem festivals**. Call the Chamber at 549-0251 for this year's events, or see their web site.

- RW

Quilt Show

OR - SUNR

Sunriver Resort

Sunriver Airport

Town	Sunriver **ID** S21 (S2-one)
Coord	N-43-52.58, W-121-27.18
Elev	4164 feet
Runway	18rt - 36, 5455 x 70' asphalt
Freq	CTAF-122.8, VASI-122.8-5x
Chart	Klamath Falls sectional, L1, Lo1

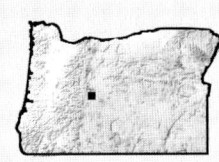

CAUTION: Airport information not for navigational use. Trees Runway 36. In winter, runway lights may be covered with snow.

> *Sunriver offers a full-featured mountain resort of luxury, golf, and access to the mountains for hiking, backpacking, or skiing. $160-320*
> *Resort: 800-547-3922, www.sunriver-resort.com*

Airport Description

The runway is paved and in good condition. The runway is kept open (plowed, if necessary) year round. Weather conditions enable the airport to be open 320 days per year for VFR traffic. For noise abatement, pattern activity should be to the west of the airport. UNICOM is active.

Services, Transportation, and Lodging

Airport phone: 541-593-4603. Fuel: 100 and Jet. Tiedown is $7 for a single. Depending on the time of year, tiedown ropes may be a scarce commodity. In summer, up to 170 planes can be accommodated in a nearby field. Cold-weather aids are available: heated hangar, engine preheat, battery jumps, and glycol treatment. Airplane rental and instruction are available.

The airport sits just a half-mile west of Sunriver Lodge. Transportation to the lodge is easily obtained by calling the lodge on a direct-connect phone. Hertz, 541-388-1535, will drop a rental car at the airport at the airport for an extra $20 fee. With advance notice, Enterprise, 383-1717, will pickup a renter from the airport.

If your plans include camping or cross-country skiing in the adjacent Deschutes National Forest, you may walk or ski directly from the airport to your destination, as described below. The lodge provides free rides to the Mount Bachelor ski area three times per day.

Transportation to prime backpacking areas in summer becomes less convenient and more expensive. I was able to uncover two options: The least creative is to simply rent a car. A second alternative is to hire a Sunriver bell-person to chauffeur you to and from the trailheads. The latter option has the advantage of allowing you to enter and exit at different trailheads. With the help of the Sunriver Lodge desk, you can negotiate such arrangements before arrival.

Lodging and meals are available close to the airport at Sunriver Lodge. Lodge rooms cost $100-170, and suites in condo like buildings cost $160-250. Reservations: 800-547-3922 or 800-801-8765. Information: 541-593-1221.

Sunriver Lodge

Overview

Sunriver offers interesting and unique possibilities. If some individuals of your party love wilderness, while others prefer lodge and hearth, Sunriver is the prescription for a happy vacation. The Sunriver offers country-club recreation, including: **three golf courses, swimming, tennis, horseback riding, bike riding, whitewater rafting, music**, and **dancing**.

Sunriver is within 30 miles of my favorite Oregon **backpacking** haunts. The Three Sisters area offers protected mountain valleys, surrounded by accessible mountains. Non-experts can climb a Sister or other local mountain in one day. Terrain and trails are relatively easy to follow. Compared to California environments of equal scenic beauty, the density of fellow hikers is low.

Prefer the comforts of an indoor hearth and bed? Why not rent a room at the Sunriver Resort. The resort was born shortly after environmental planning had become popular in Oregon. The housing and landscape development were carefully planned and controlled to be harmonious with the environment. Homes and

rentals are all modern cedar, distributed in pseudorandom pattern between trees, streams, and golf course fairways – no neon signs, mobile homes, or trash at Sunriver. The rental units are of condo-like design with fireplaces and good views. Prices are quite reasonable in winter, either side of the holidays.

My favorite routine at Sunriver is to backpack among the Three Sisters mountains for several days and return to the lodge for an hour of **hot-tubbing** followed by a civilized meal.

Downhill and Cross-Country Skiing

Mount Bachelor, 541-382-2607, is known in the northwest for its fine **downhill ski** facilities. It also has one of the country's best groomed and patrolled Nordic ski areas. The **cross-country** facility offers varied terrain, rest cabins, and access to back-country trails. For lodging at Mount Bachelor, call 389-5900. Rooms cost $83-250.

Or, you can rent skis from Sunriver and cross-country ski right from the airport. The Deschutes National Forest is located just west of the airport on the other side of the Deschutes River. If the river is covered with ice, do not take the risk of crossing – the ice covers moving water. The following instructions will get you to the other side safely and with little extra effort.

The most convenient access to the Deschutes National Forest is a bridge, approximately 1.3 miles north of the airport. To find the bridge, walk or ski north on the road west of the airport until the next intersection. Rather than turning right or left, head straight down the bike path. With the Deschutes River at your left, continue north until you reach the bridge.

After crossing the bridge, you enter U.S. Forest Service territory. On the other side, ski- and foot-trails branch off in numerous directions. Pick a trail and follow it. The forest is beautiful in any direction.

© 2001 by **ALTA** Research

Sunriver Resort

OR - TILL

Tillamook

Tillamook Airport

Town	Tillamook	**ID**	S47
Coord	N-45-25.12, W-123-48.82		
Elev	35 feet		
Runway	13 - 31, 4990 x 100' asphalt		
	1 - 19rt, 2900 x 100' asphalt		
Freq	CTAF-122.8		
Charts	Seattle sectional		

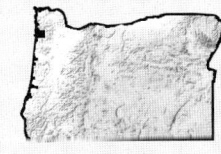

CAUTION: Airport information not for navigational use. Birds, helicopters, blimps.

Tillamook has the largest remaining wood hangar, air museum, and café on field. Cheese factories and scenic coast to explore in the area. $80-200 Museum: 503-842-1130, www.tillamookair.com

Airport Description

Photo from museum web site

Tillamook has a multi-runway airport, which originally served as a U.S. Naval Air Station in World War II. The site sits in a basin, surrounded by hills. The airport's dual runways and huge hangar are easy to spot. The hangar has "AIR MUSEUM" painted on the side. Look for the air field 6 miles east of the coast and 3 miles south of town.

Airport Services, Transportation, and Lodging

Airport phone: 503-842-7152. Fuel: 100LL and Jet. The FBO has a lounge, pilot's supplies, and a car to rent. The Tillamook Naval Air Station Museum and restaurant are on site. You can taxi right up to the museum.

E&E Auto Rental, 503-842-7802, provides rental cars for $25 per day plus a $10 airport delivery fee. North Coast Taxi is 842-4141. The Shilo Inn picks up from the Airport.

Features
- **B** Full breakfast
- **C** Continental
- **H** Historic
- **J** Jacuzzi
- **P** Pets OK
- **R** Restaurant
- **S** Swim pool
- **T** Transportation

Tillamook **OR - TILL**

The following lodging is located in or close to Tillamook. Nearby coastal towns, such as Garibaldi, offer further options in places to stay.

Best Western Inn	**CJS**	$85-128	800-299-4817
Mar Clair	**JPRS**	$47-78	800-331-6857
Red Apple Inn	**PR**	$50-100	503-842-7511
Shilo Inn	**JPRT**	$74-138	503-842-7971
Western Royal	**JP**	$55-95	503-842-8844

Tillamook Naval Air Station Museum

The Air Museum is a great destination for a day visit, and therefore a good match for residents of Oregon or Washington. You can enjoy the historic air museum, and finish the tour with a ritualistic $100 hamburger at the museum's café.

Air Base Café

The museum's hangar remains as the largest free-span wood building in the world. The area covered by the structure is equivalent to the area of 6 football fields – 1072 feet long, 296 feet wide, and over 15 stories high.

It all began in 1942 when the U.S. Navy began construction of 17 wooden hangars to house the K-class blimps for antisubmarine patrol and convoy escort along the coast. Two of the 17 were built at Tillamook.

Hangar B, the remaining hangar, was rushed to completion in 1943. Hangar A followed, and was amazingly constructed in only 30 days.

A picture gallery in the museum shows stages of construction of the hangars – quite interesting. Significant wood, mud, and manpower were involved in the project. In spite of the complexity and haste, no serious injuries or deaths have been attributed to the project.

According to museum literature, Squadron ZP 33 was stationed at Tillamook with 8 K-ships. The K-ships were 251 feet long and filled with 425,000 cu.ft. of helium. With a range of 2,000 miles and an ability to stay aloft for up to 3 days, they were well suited for their mission.

Tillamook Naval Air Station was decommissioned in 1948. Hangar A was tragically destroyed in a 1992 fire. Since 1994, the remaining hangar has been home to one of the finest collections of privately owned WWII flying aircraft in the nation.

Approximately three dozen aircraft are on display. The collection includes a significant number of classics, like the P-51 Mustang and Spitfire, as well as some antique odd-ball craft. The latest additions are a F14, PV2 Harpoon, and P2V Neptune.

The museum is open 7 days a week, 10:00-17:00. The price of admission is $9.50, and less for seniors and children. Lunch is served every day at the Air Base Café, which is open until 16:00.

Their theater features 45-minute videos of warbirds and aviation history. The Exhibit Hall has interesting photos, uniforms, and military gear. For more information, call 503-842-1130 or see the web site at *www.tillamookair.com* .

While you cannot take any of the aircraft in the museum for a ride, Tillamook Air Tours offers the next best thing. Owner Larry Stoffel provides rides along the coast in his own vintage airplanes. His Spirit of Tillamook is a fully restored 1942 "gull wing" Stinson Reliant V-77 with a 300 HP Lycoming engine. This aircraft was used in World War II for V.I.P. transport, and can carry up to five passengers.

The most recent addition to Larry's line is a 1928 Travel Air 4000 open cockpit biplane. The aircraft carries two passengers in the front seat. I can't think of a better way to see the coast.

Larry Stoffel offers rides for two in his Travel Air 4000

Tillamook OR - TILL

Prices start at $25 per person, two passengers required. Several tour packages are available, or roll your own. Whale watching is an option when the whales are running. Tillamook Air Tours is located next to the Blimp Hangar and Naval Air Station Museum. For information and reservations for a flight, call 503-842-1942.

Tillamook Area

With the convenience of a rental car, you can visit Tillamook's other offerings. Probably the most famous is the **Tillamook Cheese Factory**, 503-842-4481. Here, you can see how cheese is made in a large factory, then finish the tour with a Tillamook ice cream cone. Visit the **Blue Heron Cheese Company** to see a small-scale operation, 800 275 0638. The **WD Sausage Factory** also welcomes visitors, 503-842-2622.

Tillamook Cheese Factory

The **Pioneer Museum**, 503-842-4553, has local historical items of interest. Nine miles west, you can visit the **Cape Meares Light House**, 842-5270.

The Oregon coast has **scenic drives** north and south, and the coastal towns are fun to visit.

- RW

South of Tillamook Bay   **North of Tillamook Bay**

Photos: Larry Stoffel

© 2001 by **ALTA** Research

OR - TROU

McMenamins at Troutdale

Troutdale Airport

Town	Portland-Troutdale **ID** TTD
Coord	N-45-32.96, W-122-24.08
Elev	39 feet
Runway	7 - 25, 5400 x 150' asphalt
Freq	Twr-120.9, Lights-3x5x7x
	Gnd-121.8, Apc-133.0
Charts	Seattle sectional, L1, Lo1

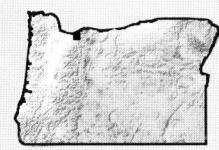

CAUTION: Airport information not for navigational use. Use Portland approach. Banner towing, helicopter ops, flight training. Waterfowl, high bluff to SE, powerlines across river to N.

> *McMenamins Edgefield, a beer-drinker's paradise, B&B, pub, brewery, winery, restaurant, and movie theater. Historic restored buildings, and every interior wall is decked in new-age art. Eat, drink, watch a movie, and sleep. $100-200*
> *McMenamins: 503-669-8610, www.mcmenamins*

Troutdale Airport

Troutdale is a tower-controlled airport with instrument approach that serves general aviation east of Portland International. The airport is often listed under "Portland" in pilot's guides. Contact Portland Approach when nearing Troutdale. Lights are pilot controlled.

Services and Transportation

Airport phones: 503-693-1963, 503-661-1044. Fuel: 100LL and Jet. Tiedown fees are $2. Hertz or Enterprise rental cars can be obtained from Premier Aircraft, 661-1044. Make rental arrangements in advance. Radio Cab, 667-1212, can deliver your party to McMenamins for about $7 plus tip.

McMenamins, the primary subject of this chapter, is located a little over one mile south of the airport. The drive from Aero West to McMenamins is 2.4 miles.

To get to McMenamins, turn right on Graham Road, the airport exit road. Follow it south past Runway 25 and under the interstate highway. You pass Columbia Gorge Factory Stores on the left. After 1.6 miles from Aero West and after crossing an overpass over tracks, turn right at the light. In a short block, bear left on Halsey and at 2.4 miles, McMenamins is to the left.

© 2001 by ALTA Research

McMenamins at Troutdale/Edgefield **OR - TROU**

McMenamins Edgefield (near Troutdale)

The McMenamin brothers, their dedicated staff, and over a dozen artists, have transformed a compound of dull government buildings into a delightful place to visit. The McMenamins campus at Edgefield (Troutdale) includes a **large-capacity B&B** operation, **fine restaurant**, **pubs**, **brewery**, **winery**, **movie theater**, **potter**, **glass blower**, and **gift shop** – a perfect evening stopover for thirsty pilots.

Originally built in 1911, the main building was opened as the County Poor Farm. The 113,160 square foot facility had 75 bedrooms. The 330 acre farm produced food for the county prison and homeless inhabitants. The concept worked until government entitlement programs made the concept of working less attractive than simply cashing a welfare check. By 1962, the facility had evolved into a nursing home.

The McMenamin brothers purchased 25 acres and the buildings in July 1990, and aggressively began restoration. The first building restored was The Power Station, which originally housed a fire station and boiler. This building now contains the pub, theater, and fourteen guest rooms. The Administration Building and Main Building followed. The McMenamin brothers have expertly converted three million dollars into an interesting environment that will stimulate the senses, put a grin on your face, and let you sleep it off in peace.

A fascinating plus is the **art**, which is literally everywhere. It is framed, on walls, on doors, on pipes, and in the brewery. Everywhere! On the second floor, I lost count at over 100 pieces. Lyle Hane is the chief artist, but there are numerous others including 14 "main artists," according to tour guide Julia Watkins. The resulting environment is a delicate balance of bizarre art and historic authenticity that somehow works.

McMenamins at Troutdale/Edgefield **OR - TROU**

The site now has 114 rooms, plus a men's hostel and women's hostel with six beds each. The cost of a stay is $23 for a single dorm bed and up. A room with breakfast for two costs $95-150. A side effect of historic authenticity is the lack of sinks, baths, and toilets in most of the rooms. To soften the inconvenience, robes are provided and the common bath areas are of top quality. Meeting other robed visitors in the hall makes a pilot feel right at home.

Family rooms and rooms with baths require advance registrations and luck to obtain. A limited number of rooms on the upper floors have sinks. Remember to request such a room if required.

When you arrive, immediately make reservations for breakfast, and also dinner if you want an elegant meal. Or, you can eat dinner at the pub. The pub offers a choice of **superb microbrews**, so don't schedule breakfast too early the next morning.

Art on every wall

Free tours enable you to see most of the **art**, **ballroom**, **restaurant**, **pub**, **brewery**, **winery**, and anything else I have failed to mention. Beer is McMenamins' primary claim to fame. They have 30-plus pubs, half of which house microbreweries. The Troutdale brewery has a twenty barrel capacity. The tour ends with a free wine tasting in the winery.

The pub has some twenty **fine beers** and **wines** on tap, many of which have the McMenamins label. My favorites are Terminator Stout and Hammerhead Ale. The pub serves excellent **pub-style food**, and is very busy on weekend nights. You can even carry your food and beer to the adjoining theater and view a **recent movie**. What a concept! The movie is free for overnight

Resident artist works in glass

guests; otherwise, one dollar. **Cigar smokers** note: the Red Shed is a must-visit spot for smoke, brew, and good conversation.

Visitors who wish to stay longer will find **Portland**, **Mount Hood**, and **Columbia River** attractions within driving distance of McMenamin's and the airport. Scenic drives along the Columbia take you through the **Columbia Gorge** area, past **Multnomah Falls** and to **Hood River**.

Mt. Saint Helens, the **volcano** that blew it's top in 1982, is another interesting place to visit. The Forest Service has done an excellent job of setting up visitors' centers. Schedule a full day for a visit to the blast area, located a good drive to the northwest.

Resident artist details a pottery piece

Oregon

Wallowa Recreation Area

OR - WALL

Enterprise Municipal Airport

Town	Enterprise	**ID**	8S4
Coord	N-45-25.49, W-117-15.89		
Elev	3957 feet, SE end high, 2%		
Runway	12 - 30, 2833 x 50' asphalt		
Freq	CTAF-122.8 unmonitored		
Chart	Seattle sectional, Lo2		

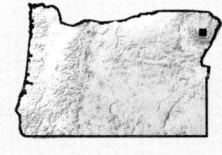

CAUTION: Airport information not for navigational use. No line of sight from runway ends.

> *The Wallowa Recreation Area offers all-around outdoor recreation: wilderness back country, as well as civilized outdoor family activities near Lake Wallowa. $75-160*
> *800-585-4121, www.wallowacountychamber.com*

Airport Description

The paved runway at Enterprise is easy to spot. The airport is elevated a couple hundred feet at the eastern edge of town. The runway is plowed in winter. Beacon and runway lights operate at night. Take care with night operations since VOR signals are blocked by the range of tall mountains seven miles southwest.

Tiedown space is ample and free. Tiedowns on inclined areas require the usual precautions.

Airport Services, Transportation, and Lodging

Phones: Spence Air, 541-426-3288. Fuel: 100 octane. If the airport is unattended, fuel is a quick phone call away. Town center is within easy walking distance. Spence Air has a courtesy car for short runs.

Camping is permitted at the airport; this may be necessary as lodging is often filled to capacity in summer. The airport bathroom is unlocked 24 hours per day.

Rental cars are available from Summit Ford, 541-426-4574, or Courtney Motors, 426-3167. Both businesses have the same owner. Rates start at $30 per day and $.30 per mile. Milligan Motors may also have rentals available, 541-426-3128. Make arrangements in advance for weekend pickup and return.

© 2001 by ALTA Research

Wallowa Recreation Area, Enterprise **OR - WALL**

Although Joseph Airport has no services, it is a worthy alternate. The airport is closer to the Wallowa recreation area and the runway has been recently improved. It is one mile from Joseph, and local merchants are generous with transportation.

Motels in Enterprise fill to capacity during tourist season; reserve your room in advance. Enterprise motels are within a 15 minute walk of the airport.

Features
- **B** Full breakfast
- **C** Continental
- **H** Historic
- **J** Jacuzzi
- **P** Pets OK
- **R** Restaurant
- **S** Swim pool
- **T** Transportation

Enterprise
Best Western Rama	JPS	$74-170	888-726-2446
Country Inn	T	$45-60	541-426-4986
George Hyatt House B&B	BHT	$100	541-426-0241
Lozier Country Loft D&B	BT	$90-130	541-426-3271
Perkins House B&B	BHPT	$70-95	541-426-3938
Ponderosa Motel	PT	$59-66	541-426-3186
Wilderness Inn	PT	$50-69	541-426-4535

Joseph
Chandler's B&B	BJT	$65-85	800-452-3781
Dragon Meadows B&B	BHT	$80	541-432-1027
Indian Lodge Motel	PT	$37-48	541-432-2651
Whitetail Farm B&B	B	$85	541-432-1630

Wallowa Lake
Eagle Cap Chalets	JST	$55-137	541-432-4704
Flying Arrow Resort	JST	$58-150	541-432-2951
Matterhorn Swiss Village	T	$48-116	541-432-4071
Tamarack Pines Inn B&B	BJT	$90	541-432-2920
Wallowa Lake Lodge	HR	$92-195	541-432-9821

The owners of Lozier Country Loft B&B also operate a **hot air balloon** business. The Tamarack Pines B&B has fishing right on the premises. The Wallowa Lake Lodge is a grand old resort,

© 2001 by ALTA Research

close to the lake with fine dining. The Wallowa Lake State Park, 541-432-4185, has a wonderful campground with easy access to the lake. Pacific Power has provided an inviting camp area with tall trees and streams. It's free, but you get to listen to the whine of their Pelton water wheel in the background.

Joseph

While Enterprise is the commercial center for the area, nearby Joseph has become the favorite of recreational visitors. Joseph airport is handicapped by lack of services and greater distance from town. Local businesses compensate by providing complimentary transportation. The Manuel Museum, for example, offers free rides to any group that visits the museum. Further, they provide inexpensive transportation for pilots to other points of interest.

A bank robbery waiting to happen, Joseph

Wallowa Recreation Area, Enterprise   **OR - WALL**

Joseph features a **museum**, **bronze foundry**, **galleries**, **cowboy bars**, and **staged bank robberies.**

If you visit Joseph, don't miss the **Manuel Museum**, a.k.a. **Nez Perce Crossing**, 541-432-7235. David and Lee Manuel have built an amazing art and American history empire in Joseph. David creates nationally acclaimed **bronze sculpture** with a pioneer and Native American theme. He has supplemented his interest in Native American culture by assembling a museum of beautiful Nez Perce artifacts.

The **art gallery** and **Indian history museum** are housed in a magnificent 14,000 square foot structure built of logs. Lee keeps the business humming with promotional presentations. Ask her about their **art classes**, **seminars**, **field trips**, **area bus tours**, and local **foundry tours**.

Manuel Museum

Steens Wilderness Adventures offers **float trips** and other outdoor excursions, 541-432-5315. Eagle Cap Pack Station is in the Wallowa Wilderness and specializes in **hourly horseback rides**, **pack trips**, and **hunting and fishing excursions**, 800-681-6222. The USFS Visitor's Center is at 541-426-5546.

Hell's Canyon Adventures, about 1.5 hours from the area, gives **jet boat tours** and float trips on the Snake River. Jet boat prices range from $40 to $100 for two- to six-hour tours, and a one-day **raft trip** costs $135. Call Bret Armacost at 800-422-3568 for current prices and details.

Wallowa Lake

Wallowa Lake sparkles like a jewel in northeastern Oregon. Cradled at the lake's southernmost tip is a recreational center that can provide days of outdoor entertainment. Options include

Deer visit the Wallowa Lake Lodge

Wallowa Recreation Area, Enterprise **OR - WALL**

exceptional **hiking, boating, horseback riding, tram rides, camping,** and quality meals at the Wallowa Lake Lodge.

Wallowa Lake village is eleven miles southwest of Enterprise. From Enterprise, head southeast on Highway 82 to Joseph. Follow signs through Joseph to Wallowa Lake. When you reach the Wallowa Lake Lodge, bear left for most of the local action. You pass horse rentals, the tramway, galleries, and several resorts before the road dead-ends at the base of the mountains in less than a mile. Trails radiate from the Pelton wheel power station at the end.

Bear right at the lodge-intersection to get to Wallowa Lake State Park, 541-432-4185, and marina, 432-9115. The marina rents all sizes and shapes, paddle **boats** to powered models.

The **Wallowa Lake Tramway** features the steepest vertical rise in the U.S. The 15 minute ride (each way) costs $6 for children and $10 for adults, round trip. Cheaper than flying. The ride takes you up 3700 feet to the summit of 8150. At the top, you have the run of two miles of trails with 360 degree views and a snack bar. The lift is open late May through late September (last day of Alpenfest). On either side of tourist season, check for part-time operation by calling 541-432-5331.

Events

The big event at Wallowa Lake is **Alpenfest** in late September, the third weekend after Labor Day. Edelweiss Hall, next to the tramway, is transformed into a bit of Old Switzerland with lavish decorations, yodeling, dancing, alpine horn blowing, and serious beer drinking. Activities in town peak during these days of celebration. Make reservations early. For information, call the Wallowa Lake Tourist Committee at 541-432-4704.

© 2001 by **ALTA** Research

The big event in Joseph is **Chief Joseph Days**, a premier **rodeo** that occurs the last weekend of July. The celebration includes four rodeo shows and two parades. In summer, Enterprise and neighboring towns host an event almost every weekend. You can get a list by calling the Wallowa County information number, 800-585-4121, or Joseph Chamber, 541-432-1015. A sampling of local events and festivals follows:

> Crazy Daze
> Hot Air Balloon Festival
> Oregon Mountain Car Cruise
> Jazz at the Lake
> Chief Joseph Days
> Hells Canyon Mule Days
> Bear or Rattle Snake Feed
> Alpenfest

- RW

Wallowa Lake country

Washington

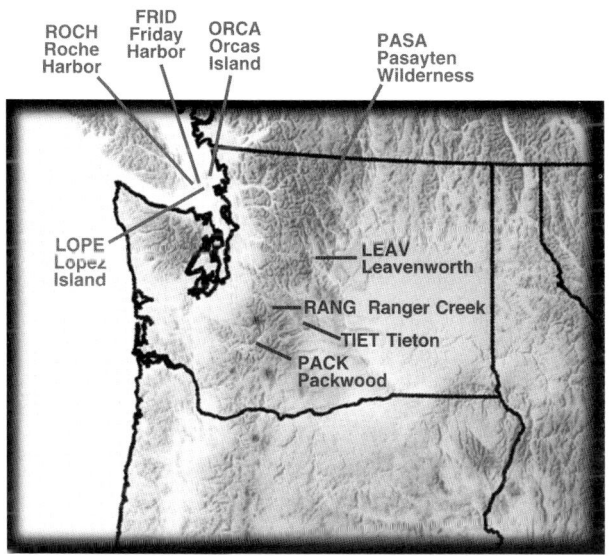

Washington Aeronautics

206-764-4131

Washington

© 2001 by ALTA Research

WA - FRID

Friday Harbor

Friday Harbor Airport

Town	Friday Harbor **ID** FHR
Coord	N-48-31.32, W-123-01.46
Elev	113 feet
Runway	16rt - 34rt, 3400 x 75' asph.
Freq	CTAF-128.25, ATIS-134.15
	Lights-128.25-3x5x7x
Charts	Seattle sectional, low-alt L1, Lo1

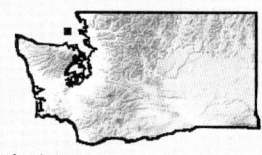

CAUTION: *Airport information not for navigational use. Active traffic. Deer. Occasional heavy crosswinds. Town just north; use noise abatement procedures.*

> *Friday Harbor sits on a picturesque island, the commercial center for the laid-back San Juan islands. Visitors enjoy B&Bs, fishing, ferry rides, bicycling, and strolls by the sea. $110-300*
> *Info: 360-468-3663, CofC: 360-378-5240,*
> *www.friday-harbor.net, www.sanjuanisland.org,*
> *www.guidetosanjuans.com*

Airport Description

The public airport is a large asphalt runway at the eastern side of San Juan Island, just south of Friday Harbor. Take care not to land at the 1500-foot strip 1.3 miles west of the airport. Although landing at Friday Harbor can be safe and straightforward, be alert for traffic and excessive crosswinds.

Weather is better at the islands than on the mainland. While Seattle is veiled in marginal VFR greyness, the islands often enjoy sun and scattered clouds. Contact Whidbey Island Naval Air Station on 118.2 for flight-following over water. Runway lights are controlled by clicking 128.25. The United Parcel Service pilots at Friday Harbor are a good source of tips for flying the islands.

Airport Services

Airport phone: 360-378-4724; FBO: 378-2640. Fuel: 100 octane, available 24 hours with credit card. The airport is attended during normal island working hours.

Visitors may tiedown on the paved apron near the fuel pumps or north of the runway. The tiedown fee is high. Tiedown can be scarce in summer, especially during the jazz festival, typically the last weekend in July. Camping is no longer allowed.

© 2001 by ALTA Research

Friday Harbor **WA - FRID**

Stores at the north end of the airport serve fly-in grocery shoppers. Additional shopping and restaurants are within a ten-minute walk. Several blocks more, and you are at the wharf.

Transportation

With the town and harbor so close to the airport, motorized transportation is not necessary for short visits. To get to the town and harbor, walk north to the end of the airport, turn right on Spring street, and walk for another five minutes into town.

Just before entering the central part of town, you pass the Inns at Friday Harbor on the left. The Inns rents cars for $50 per day, plus extra for mileage. Before they will rent, the Inns requires the following: your insurance company's name, proof of coverage, and your agent's name and phone number. Call 800-752-5752. Mopeds can be rented from Suzie's Mopeds in town, 360-378-5244. Rates are $16 per hour or $48 per day.

Island Bicycles at 360-378-4941 rents all sizes and flavors of bicycles. You will find their shop one block northwest of Spring Street, near the corner of First and West Streets. Island Bicycles offers a variety of rental plans.

Due to the expense of interisland automobile travel, islanders use hitch hiking like mainlanders use a bus or taxi. Don't feel shy about asking for a ride or extending your thumb. Most inns provide pickup from the airport; verify when making reservations. If all else fails, call San Juan Taxi at 360-378-3550.

The best bargain in transportation is the Washington State Ferry system, which is free to all interisland foot passengers. Do not pass up a delightful, free, round-trip tour of the islands. Pick up a timetable from the dock when you arrive in Friday Harbor.

© 2001 by ALTA Research

Lodging and Meals

The 132-unit Inns at Friday Harbor is the largest facility at the east end of the island. The Inns provides pickup service, but the seven-minute walk to its location at 410 Spring Street is faster. The Inns is a center for car rentals, **tours**, and information. Rooms are standard motel fare: clean, with medium-quality decor. Amenities include TV, access to a central Jacuzzi, indoor pool, and exercise room.

The Snug Harbor Marina Resort caters to **fishermen** and **scuba** divers. It also offers **kayak tours**, **whale watching**, and **sunset boat tours**. The Resort is located at 2371 Mitchell Bay Road at the northwest end of the island.

Island B&B establishments book to full capacity in summer months; always make reservations in advance. The most unique is the Wharfside, a 60-foot boat moored in the harbor. The

Ferry arrival at Friday Harbor

Friday Harbor   **WA - FRID**

Duffy House has a notably scenic location on Griffin Bay. In the following list, inns marked with an asterisk are within four blocks of the wharf.

Blair House B&B *	JB	$95-180	800-899-3030
Discovery Inn	J	$58-134	800-VAC-ISLE
Duffy House B&B	BHT	$113-135	800-972-2089
Hillside House B&B	BJ	$75-270	800-232-4730
Inns at Friday Harbor	JPST	$75-195	800-752-5752
Lakedale Resort	CS	$80-340	800-617-2267
The Meadows B&B	BT	$90-155	360-378-4004
Olympic Lights B&B	BH	$90-145	360-378-3186
Roche Harbor Resort	HRS	$75-300	800-451-8910
San Juan Inn *	CHJP	$73-205	800-742-8210
Snug Harbor Resort	PR	$102-226	360-378-4702
Tucker House B&B *	BHJPT	$135-243	800-965-0123
Wharfside (boat) B&B *	BT	$146-178	800-899-3030

Lakedale Resort is 5 miles northwest of the airport off Roche Harbor Road. Tent camping costs $7-8 for two people without a car, and $20-25 with a car. **Swimming**, **boating**, **fishing**, and showers are available for a fee. The Peddle Inn is a rustic **bicycler's campground**, five miles south of the airport on False Bay Road, 378-3049.

Features
B Full breakfast
C Continental
H Historic
J Jacuzzi
P Pets OK
R Restaurant
S Swim pool
T Transportation

Roche Harbor Resort is at the other end of the island. See **WA-ROCH**.

Friday Harbor has a number of delightful restaurants. The Springtree Cafe offers good food at fair prices. Try the Springtree for breakfast or dinner, and the Cannery House on First Street for lunch. Fresh seafood is the speciality of several restaurants in Friday Harbor.

© 2001 by ALTA Research

Friday Harbor and San Juan Island

San Juan Island is the county's busiest island, a center of commerce for the 172 picturesque islands that make up the San Juan Archipelago. Friday Harbor is the county seat and its largest town. Yet even during the bustling tourist season, Friday Harbor maintains an easygoing ambiance, characteristic of the islands.

San Juan Island is lightly populated. Terrain is gently rolling with occasional outcroppings of rock and hills. Roads glide past pastures and forests, drop to taste a salty bay, and rise with the bluffs to catch a marine panorama. The island country offers an idyllic break from the pace and population to which most of us have become accustomed.

Upon returning from a day in the country, Friday Harbor's warm hum of activity primes us for an evening of good food and pleasant conversation. Streets near the harbor host restaurants

and shops a cut above the norm. Pilots appreciate that most of the town's offerings are within easy walking distance.

What to Do

San Juan Island is a favorite of avid **bicyclists**. The scenery is varied and pleasant. With few exceptions, hills are gentle and negotiable by the casual bicyclist. A ride to the south tip and back requires two to four hours, including breaks and picnic. (See Transportation above.) Prefer simply to sit and enjoy? Take a **tour** with San Juan Transit from the Inns at Friday Harbor, 360-378-8887.

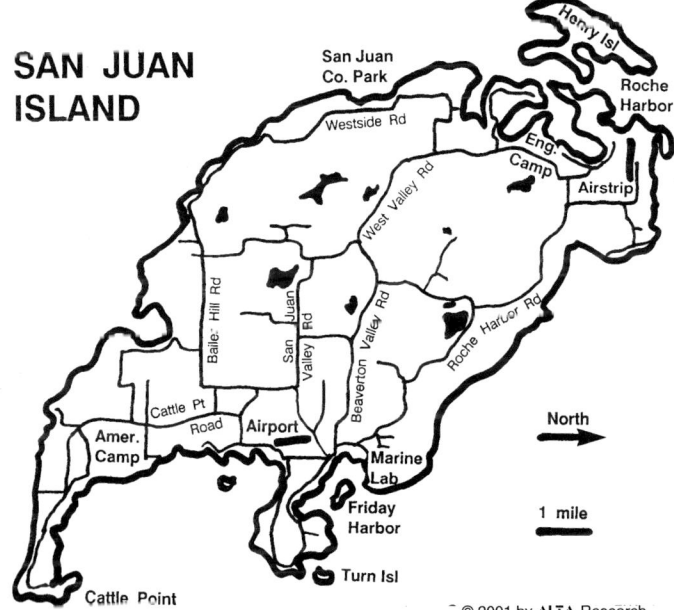

The island offers a range of **boating** options. **Rentals**, **fishing**, **sight-seeing**, and **diving** are available from San Juan Boat Tours, 800-232-6722. Lakedale Campground has small boats for rent, 360-378-2350. Snug Harbor Marina Resort, 360-378-4612, operates a full-service marina. Snug Harbor books **Whale Watch Charters**. They also book for Buffalo Works Nash Brothers who provide **fishing charters** and **wildlife cruises**. Western Prince Cruises specializes in wildlife cruises, 800-757-6722.

You can rent a boat at Friday Harbor and row or motor two miles to Turn Island **wildlife refuge**. The island is a 35-acre park with 10 campsites on the southwest end. Bring your own water.

Remember that Washington State offers the most cost-effective option for boat rides: free interisland **ferry rides** for walk-on passengers. Pickup a timetable at the dock, call 800-84-FERRY for schedules, or see *www.wsdot.gov/ferries* .

Residents claim that San Juan County has more **museums** per capita than any other county in the United States. Friday Harbor's **Whale Museum** is on First Street and the San Juan Island **Historical Museum** is one-half mile south of the dock on Price Street. The University of Washington **Marine Laboratories,** one mile north of town, are open to the public between 2:00 PM and 4:00 during the summer.

Friday Harbor has the only **movie theater** on the islands. The single-screen theater provides a cozy old-time environment at reasonable prices. It shows up to 3 different movies per day.

The biggest and most lively event in the San Juans has been the **Dixieland Jazz Festival,** typically the last weekend in July. Unfortunately, the festival may be discontinued. Call the Info or CofC number for latest details.

Also see **WA-ROCH**, also on San Juan Island.

- RW

© 2001 by ALTA Research

Leavenworth

WA - LEAV

Cashmere - Dryden Airport

Town	Cashmere **ID** 8S2
Coord	N47-30.89, W120-29.09
Elev	853 ft, W end high, 1.3%
Runway	7 - 25, 1850 x 50' asphalt
Freq	CTAF-122.9,
	Lights-121.7 - 3xON, 5xOFF

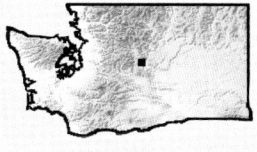

CAUTION: Airport information not for navigational use. Tall mountains in all quadrants. Towers and trees nearby. Helicopter & ag ops.

Photo by Sandy Campbell

> *Leavenworth is a picture-postcard Bavarian town, seemingly transplanted from southern Germany, piece by piece. Experience colors, tastes, and hospitality you won't forget. $100-200*
> *509-548-5807, www.leavenworth.org, & ...com*

Airport Description

Cashmere Airport is nearly located in the center of town; if you can find Cashmere, you've found the airport. Note that the runway is short and you will need capability to climb with authority. If unable to safely meet these criteria, consider landing at an alternate, like Wenatchee. Due to runway grade, land on Runway 25 and depart on 7 unless winds favor otherwise.

Services, Transportation, and Local Lodging

Airport phones: 509-782-5561, cell 421-1768, 782-3256, cell 679-3485. Fuel, emergencies only. Call the above numbers for fuel or access to a courtesy car. The town of Cashmere and bus stop are just a short walk from the airport. You can ride the "Link" bus to get to Leavenworth. There are no rental cars in town, but the bus (800-851-LINK, functional locally only) is a cost-effective way to get to Leavenworth. The ride to Leavenworth takes about 20 minutes. Local taxi service is provided by Mountain View Taxi, 782-3346.

In Cashmire, you can camp on the air field and use the facilities in the pilot's lounge, or stay at the Village Inn Motel, 800-793-3522. The motel charges $49-63, does not pickup from the airport, and does not allow pets.

© 2001 by ALTA Research

Leavenworth **WA - LEAV**

Leavenworth Lodging

Leavenworth has over 100 places to stay. The following inns are close to central Leavenworth. Given advance notice, they claim they will pickup from Cashmere airport. Visit the Leavenworth web site for a more comprehensive list, or call the Bavarian Bedfinders at 800-323-2920. Ask for a place that meets your needs – from cabins or condos, to B&Bs.

Features
- **B** Full breakfast
- **C** Continental
- **H** Historic
- **J** Jacuzzi
- **P** Pets OK
- **R** Restaurant
- **S** Swim pool
- **T** Transportation

Name	Features	Price	Phone
Bash Garten B&B	BJT	$110-140	800-535-0069
Der Bear-Varian	JT	$143-154	800-548-1858
Der Ritterhof	CJPT	$84-168	800-255-5845
Enzian Inn	BJSPT	$84-168	800-223-8511
Haus Loreli B&B	BJT	$106-145	800-514-8868
Pension Anna B&B	BJT	$89-232	800-509-ANNA
Tumwater Mtn. Inn B&B	BJT	$95-106	509-782-0708

When making arrangements, remember to ask about the establishment's willingness to pickup from Cashmere, distance from town center, and availability of courtesy bikes.

Leavenworth

Like a number of other attractive tourist towns in the Flyer's Recreation Guides, Leavenworth began its existence in the late 1800s as a rowdy frontier town. A sawmill provided the economic boost that placed Leavenworth on the map. The mill received logs that were floated down the Icicle and Wenatchee rivers from logging camps upstream. The Great Northern Railroad soon followed, adding a switch-yard that seemed to insure the small town's chances of survival.

Fruit farming sustained Leavenworth's economy after the mill closed in the mid-1920s, but the decline became serious

© 2001 by **ALTA** Research

when the railroad was rerouted through the Chunstick Valley in 1928. By 1960, over twenty storefronts were empty; and unless something were done, Leavenworth would soon become a ghost town.

Local residents rose to the occasion. By working together and with help from the University of Washington, they decided to place their bets on tourism. Inspired by the Danish town of Solvang in California, the merchants agreed to convert Leavenworth into a Bavarian theme town.

They did a beautiful job. Today, the town has over one hundred places to stay, two dozen places to eat, and a variety of outdoor recreation. You can **hike**, **climb mountains**, **ride bikes**, **play golf**, or run the Wenatchee with one of ten professional **whitewater rafting** outfits.

Although there is plenty to do in the area, most of the activity seems to center around **food**. With their local brewery, pastry, sausages, and other culinary delights at every turn, how can a pilot resist such temptations? Between snacks, you can browse the stores that are filled with German art and gifts.

Leavenworth has a major **festival** almost every month of the year. Most include plenty of music, food, dance, and costumes. Call the Chamber of Commerce for a schedule.

John and Sandy Campbell fly their Cessna 182 from their ranch in Montana to visit Leavenworth whenever they feel due for a special treat. Sandy's impressions follow:

"Cashmere is located within the heart of apple country, only 21 minutes to Wenatchee or 20 minutes to Leavenworth. The greatest discovery for us was the *free* Link bus service that goes to fifteen towns and several associated air strips in the area. One quarter of 1% sales tax goes to fund this excellent service. Locally, call Link at 800-851-LINK for the schedule or pick one up at the Cashmere pilot's lounge. The bus stop is three blocks from the Cashmere Airport.

"Cashmere is small, with a few shops and some very good restaurants: The Pewter Pot, 509-782-2036, and the Aplets and Cotlets Factory, 782-2191, with free 15-minute tours and samples of their dried fruit yummies, for example.

"After the tour, we quickly caught the bus to Leavenworth. Leavenworth is a colorful Bavarian town, complete with authentic costumes, roving accordion players, Bavarian polka music through every sound system in town, flowers galore, and German food that will guarantee a satisfied tummy. There are also fine Mexican, Italian, Greek, and French restaurants for lots of variety.

"We discovered an excellent place for breakfast in Leavenworth at 8th and 2nd, downstairs below the new Kenny Building. It is owned and run by a bubbly couple from Seattle.

"The **Leavenworth Brewery and Pub** was new in 1992. We checked this out thoroughly and decided we had found *it*. And, as luck would have it, our hotel, the Tyrolean Ritz Hotel was just across the street from *it*.

"There are lots of stores to shop with samples of mustard, salsas, and candies to taste. You can walk to **Waterfront Park** for a change of pace. Area entertainment includes **mushroom hunting**, **art** in the park, **golfing**, **festivals**, **rock climbing**, **hiking** in the enchanted Lakes wilderness, **mountain biking**, **rafting** on the Wenatchee River, and more. Stop at the Chamber of Commerce and pickup their calendar of events, the *Cascade Loop Travelers Guide*, and the free *Sonnenschein of Leavenworth* paper.

"The only drawback we could find in either town is that except for restaurants, everything closes at 6:00 PM. That leaves

Photo: Sandy Campbell

Leavenworth

dining, and we could only stretch our meal for so long. Going to bed early was an unexpected treat that we had not planned.

"We've been to Leavenworth many times since our first trip. Each time we plan to see other areas serviced by the Link bus. And, each time we get captivated by the beauty of Leavenworth and the oom-pah-pah music. Unfortunately, we discovered, the airport in Wenatchee is not connected to the Link. Its closest stop is one mile from the airport. Now veterans of the Link, we have learned to pack very lightly – only the essentials!"

In town, you may want to stroll to the **brewery, chocolate factory,** and **Nutcracker Museum** on Front St. Two miles outside of town is the **Leavenworth National Fish Hatchery**.

Fortunately, the Leavenworth area offers an anecdote for all those freshly accumulated calories – **outdoor recreation**:

All Rivers Adventures	*rafting*	800-743-5628
Alpine Adventures	*rafting*	509-782-7042
Cascade Playtime Rental	*snowmobile*	509-649-2444
Eagle Creek Ranch	*horse, sleigh*	800-221-RIDE
Enchanted Water Tours	*rafting*	888-723-8987
Leavenworth Outfitters	*ski, rafting*	800-347-7934
Mtn. Spr. Lodge	*horse, hay, carriage*	800-858-2276
Mtn. Spr. Lodge	*snowmobile, sleigh*	800-858-2276
Osprey Rafting Co.	*kayaks, rafting*	800-743-6269
Red Tail Canyon Farm	*hay, sleigh*	800-678-4512
River Riders, Inc.	*rafting*	800-448-RAFT

Golfers can play at the 18-hole Leavenworth public golf course, 509-548-7267. **Hikers** can obtain maps and information from the Leavenworth Ranger Station, 548-6977. There's also a variety of nearby **skiing**, **hunting**, and **fishing**. For fishing information, or if you would like to learn the art, call the School of English Fly Fishing at 763-3429.

© 2001 by ALTA Research

Photo: Sandy Campbell

WA - LOPE

Lopez Island

Lopez Island Airport

Town	Lopez Village **ID** S31
Coord	N48-28.96, W122-56.21
Elev	200 feet
Runway	16rt - 34, 2900 x 60' asph.
Freq	CTAF-128.25
Charts	Seattle sectional

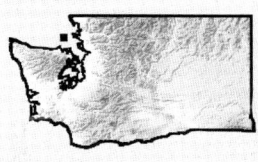

CAUTION: Airport Information not for navigational use.

Lopez is a quiet, pastoral island. It's off the beaten tourist track, an agricultural ocean-side environment with facilities for visitors. $90-150 Info: 360-468-3663, CofC: 360-468-4664, www.guidetosanjuans.com, www.lopezisland.com

Airport Description

Look for the airport on the western edge of the island. The north/south runway lies among stands of trees, with a small golf course to the east and the ocean to the west. Windsocks are near mid-field, either side of the runway.

Airport Services

Airport information: 360-468-2869, 468-2131. The airport is unattended and has no services. Phone, toilet, water, and waiting room are located near the tiedown area. Overnight camping is no longer permitted.

Transportation

Taxi: 360-468-2227. Although rental cars are not available, the islanders compensate by being courteous to hitchhikers and providing pickup services from their businesses. The Lopez Islander Inn, for example, provides guests with free pickup from the airport and as a courtesy, drives them to other destinations on the island during their stay.

Because all points of interest are within eight miles and the island is relatively flat, bicycles are a popular mode of transportation. Bring your own fold-up or rent from the Lopez Bicycle

Lopez Island **WA - LOPE**

Kayak Works, just south of Lopez Village. Call them at 360-468-2847 and ask if they can pick you up from the airport, or ask them to leave bikes at the airport.

Lodging and Meals

The Lopez Islander Resort is a marine resort with pool, hot tub, and field games. It is conveniently located south of Lopez Village within walking distance of shops and good restaurants. Edenwild Inn is located in town.

Features
B Full breakfast
C Continental
H Historic
J Jacuzzi
P Pets OK
R Restaurant
S Swim pool
T Transportation

Name	Features	Price	Phone
Aleck Bay Inn B&B	BJT	$106-190	360-468-3535
Blue Fjord Cabins	T	$85-106	888-633-0401
Edenwild Inn (in town)	B	$95-185	800-606-0662
Inn at Swifts Bay B&B	BJT	$103-200	360-468-3636
Isl. House Vacation Rentals	T	$70-300	800-781-2882
Lopez Farm Cottages	BJT	$106-162	800-440-3556
Lopez Islander Resort	JSRT	$85-280	800-736-3434
MacKaye Harbor Inn B&B	BHT	$74-150	888-314-6140

The following inns accept guests at rustic extremities of the island. They typically restrict smoking, pets, and small children. The Blue Fjord Cabins, open May 1 to October 1, are positioned on the eastern edge of the island. The cedar log cabins contain kitchens and are privately set amidst trees near the water's edge. The Inn at Swifts Bay is a five-unit B&B located two miles southeast of the ferry landing. The facilities are located at the edge of a woods with access to private beach and hot tub. The MacKaye Harbor Inn B&B has five upper-story rooms, three of which have a view of the bay.

© 2001 by ALTA Research

When making reservations at any of the above inns, remember to ask for a unit with a view, and verify that transportation from and to the airport is part of the package. Alternatively, you can camp at Lopez Farm Cottages or at Spencer Spit State Park.

My favorite all-around restaurant was the Bay Cafe in Lopez Village. Playing close a close second was Buckey's Grille. Prices were reasonable.

What to Do on Lopez Island

Ask any tourist for his impression of Lopez Island. Odds are, he will respond with adjectives like peaceful, pastoral, or even boring. City visitors have been known to roll off the ferry, take a quick trip around the island, and catch the next departing ferry. Despite its modest appearance, Lopez Island somehow touched my heart within minutes after stepping onto solid ground.

The island has several day-use recreation sites, three overnight parks, and a small village. Fly-in visitors with only several hours of time should consider visiting the **Shark Reef Recreation Site** just south of the airport. Walk two miles due south on Shark Reef Road. After entering the park, a short hike brings you through the park's old-growth forest to the coast. With the aid of binoculars, you can enjoy harbor seals and sea lions sunning themselves on rocks.

Golf enthusiasts may enjoy visiting the Lopez Golf Club, located within easy walking distance of the airport. Fees are reasonable and clubs are available for rent. Call 360-468-2679 to verify someone will be there when you visit.

The island's leisurely commercial activity is centered around Lopez Village, four miles north of the airport. The distance and lack of convenient transportation from the airport can make

Lopez an awkward meal stop. Too bad, because the village has good restaurants and a yummy bakery.

Flatland **bicyclists** appreciate the ease of peddling on this small, flat island. The sites that effortlessly roll by are fields, pastures, sheep, goats, hedgerows, fishing gear, stands of forest, and rugged coast. Passersby wave; Lopez is the friendliest and most laid-back of the big four.

When the tasting room is open, **Lopez Island Vinyards** is a great bicycle stop. Call 360-468-3644 for hours.

Located at the north end west of the ferry landing, **Odlin County Park** is a great stop for **family entertainment**. Unlike most big-government parks, it includes places for children to play, as well as the standard run of **beaches** and **trails** found at other island parks. While kids create sand castles and romp in the sand, the chilly waters lure hearty swimmers.

At low tide, try your luck **digging for clams**. **Scuba divers** frequent the area between the dock and rock walls of Upright Head. If hunger strikes, you can fix your need by ordering a juicy hamburger at the snack shack near the ferry landing. **Odlin Park** is 80 acres, includes 4000 feet of shoreline, and has 30 improved **campsites**. Contact the park office at 360-468-2496 for information.

Spencer Spit State Park is the only other public park with **campsites**. Approximately half again as large as Odlin, Spencer enjoys a choice location at the northeast edge of the island. During summer, the beaches collect enough heat to render the chilly local waters just bearable for **swimming**. The sand hides butter, horse, and littleneck **clams** that reluctantly surrender to diggers at low tides. Call 360-468-2251 for further information.

- RW

© 2001 by ALTA Research

Orcas Island

Orcas Island Airport

Town	Eastsound **ID** ORS (oh-RS)
Coord	N-48-42.49, W-122-54.83
Elev	31 feet
Runway	16 - 34rt, 2900 x 60' asphalt
Freq	CTAF-128.25, Lights-128.25-5x
Charts	Seattle sectional

CAUTION: *Airport information not for navigational use. Strong crosswinds not uncommon.*

WA - ORCA

> *Orcas Island is diverse in beauty, home to artists, and host to vacationers. Of the San Juans, Orcas offers the most options, a smorgasbord of scenery, miles of hiking, camping, a variety of lodging, gourmet dining, art, boating, and fishing. $20-250*
> *Info: 360-468-3663,   CofC: 360-376-8888,*
> *www.guidetosanjuans.com,   www.orcasisland.org*

Airport Description

Look for a prominent north/south runway at the northern edge of the Orcas-Island isthmus. The airport is less than one mile north of Eastsound. The runway is lighted at night. When departing Runway 34, climb to above 500 feet and fly past the shoreline before turning.

Airport Services and Transportation

Airport phone: 360-376-5285, 376-6769. Fuel: 100LL. The FBO may be unattended while the operator is giving flying lessons. Pilots are permitted to camp at the north end of the airport. I discretely camped between the north end and the marina. This site provides a view of the ocean and access to a portable toilet used by boats at the marina. Portable toilets are also located near the airport buildings.

A walk into Eastsound, the largest town on the island, takes ten minutes. Go to the south end of the airport. Turn left, walk one block east, and turn right on the road into town. You can trim two minutes by taking the southeast diagonal path that leads from the airport to the road into town.

© 2001 by ALTA Research

Orcas Island **WA - ORCA**

Orcas Taxi provides taxi service, 360-376-TAXI. Call before departure and the taxi will be there when you land. Wildlife Cycles, 376-4708, rents bikes for $25 per day and up. Hitch hiking is exceptionally good in the islands.

Practical Rent-a-Car, 360-376-4176, rents cars at the airport for $50. Rental cars are also available at Rosario Resort for $65-90. Rosario's, 800-560-8820, provides guests with scheduled transportation to and from the airport and ferry. The ferry landing is located at the other end of the island and there are no formal connections between ferry and airport.

Lodging

Camping is permitted at the airport. The only public park camping on Orcas is at Moran State Park and a small hike-in campground at Obstruction Pass. Reservations at Moran are a must; call 800-452-5687. Doe Bay (southeast) and West Beach (west) are private facilities that cater to RV and tent campers.

The closest B&B is Kangaroo House, located one block east of the airport. Smuggler's Villa Resort is a newer facility just northeast of the airport, a good pick for fly-in families, as room rates include up to four persons. North Beach Inn is one-half mile northwest of the airport and has rustic cabins on a private beach. The Landmark Inn and Outlook Inn are located in Eastsound with views of the harbor.

Other options are less conveniently located. The restored Victorian Orcas Hotel sits proudly above the ferry landing at the southwest tip of the island. Beach Haven Resort advertises its rustic log cabins, "For those with the courage to holiday without TV and phones . . ." Beach Haven is two miles southwest of the airport and requires a minimum of seven nights stay in summer. Deer Harbor Resort has cabins with kitchenettes, some with spa, and a swimming pool.

© 2001 by ALTA Research

Rosario Resort and Spa is a prominent historical landmark that aspires to be the class resort of the San Juan Islands. Perched on a hillside in an Eastsound cove, Rosario's has kept pace with the luxuries of today, while retaining a flavor of the past. If you stay at Rosario's, take time to hike the beautiful trail to Cascade Lake, nearly 400 feet in elevation above.

At the other end of the spectrum is scenic and serene Doe Bay Village Resort. Prices for lodging are relatively low. For $24-28, you can pitch your tent at the most beautiful secluded campsites I have ever seen. The resort has the pace and aura of a spiritual retreat. Guests may sweat in a **wood-stoked sauna**, look down on the bay while soaking in **hot mineral baths** (clothing optional), or enjoy a massage on site. Non-overnight visitors pay for use of sauna and baths. **Island kayaking** by Shear Water Kayaking is a speciality of Doe Bay Village Resort.

Rosario Resort

Orcas Island **WA - ORCA**

When choosing lodging, ask about location and options for transportation. Lieber Haven Marina Resort, for example, has an awkward location at Obstruction Pass, the southeastern tip of Orcas. However, around the marina's pier are trading post, grocery store, **fishing** and **sailing charter boats**, and **Kayak rentals**. The Lodging Vacancy Hotline can help locate lodging at the last minute, 360-376-8888.

Features
B Full breakfast
C Continental
H Historic
J Jacuzzi
P Pets OK
R Restaurant
S Swim pool
T Transportation

Lodging	Features	Price	Phone
Beach Haven	H	$107-243	360-376-2288
Cabins-on-the-Point	JHPT	$162-378	360-376-4114
Cayou Cove on Orcas Isl.	BHJT	$139-265	888-596-7222
Doe Bay Village Resort	JR	$64-108	360-376-2291
Inn at Ship Bay	DIIRT	$108-211	360-376-3933
Kangaroo House (airport)	BHJT	$96-151	888-371-2175
Lieber Haven Resort		$140-146	360-376-2472
Orcas Hotel	CHJRT	$53-106	888-672-2792
Outlook Inn	HJPRT	$47-313	888-OUTLOOK
Resort at Deer Harbor	CJR	$139-430	888-376-4480
Rosario Resort	HT	$172-430	888-688-5665
Sandcastle Guest House	CT	$97-135	360-376-2337
Smuggler's Villa Resort	JST	$107-268	800-488-2097

What to See and Do

Among the San Juans, Orcas Islanders may well boast that Orcas has the most of everything: more area, more shore line, more protected waters, and taller mountains. Unlike the other islands whose commercial resources attracted early settlers, Orcas earned early fame as a vacation island. Recreation became the mainstay of the island. In later years, Orcas attracted artists and crafts persons who now supply the island's many galleries. The community actively supports a **local theater,** 360-376-2281, one-quarter mile east of the airport.

© 2001 by ALTA Research

Moran State Park spreads 4934 acres through the central portion of the island's eastern lobe. Besides its **lush forests** and **scenic lakes**, the park's big draw is the view from 2407-foot **Mt. Constitution**. Thirty-five miles of interconnecting **trails** radiate from **camping areas** within park boundaries. Avid hikers can easily spend several days exploring the varied terrain. Call 800-452-5687 for reservations and mail confirmation at least two weeks in advance if you plan to stay at one of the park's 151 campsites. Cost is $10 per campsite.

Obstruction Pass is a scenic treasure. Located at the southern tip of the eastern lobe, it offers views to savor, a **hike-in campground**, and the rustic luxury of the Licber Haven Resort. To get there, follow Horseshoe Highway East from Eastsound to Olga. Just before entering Olga, stop at the Orcas Island Artworks. Enjoy the **local art** and treat yourself to a snack in their quality restaurant. Follow the road toward Doe Bay for 1.5 miles

Lieber Haven Marina Resort

and turn south on Obstruction Pass Road. One mile after this intersection, turn right on the gravel road that heads to the park.

After enjoying the **trails and views**, head toward Doe Bay Village for a soak in the resort's **heated mineral springs**. If you have made reservations at one of the island's fine restaurants, you can end your day with a satisfying meal.

Consider **renting a boat** and visiting local islands. No boating proficiency is necessary. The 172 San Juan Islands offer visual experiences that can only be appreciated from sea. **Sucia Island** is the largest, a 2.5-mile (20 minute) ride from the marina near the airport. The island has a most interesting perimeter, with mile-long rocky fingers protecting shallow inlets. The Sucias are a collection of eleven protected islands. The Park Service, 800-233-0321 maintains 51 **campsites** on the main island with picnic tables, shelters, fireplaces, water, dock, and pit toilets.

Doe Bay Village Resort mineral springs

Patos Island State Park is six miles across the water northwest of the airport. It has a lighthouse on its western tip and a three-site campground in a southwestern cove. The 209 acre island has toilets and tables, but *no water*. **Matia Island** State Park is located three miles northwest of the airport. Said to be one of the prettiest islands, it is often passed over by sleep-in boaters due to its limited overnight moorings. The island has six campsites with the usual amenities, except *no water*. Only five acres are open to campers; the other 140 serve as a wildlife refuge.

Trophy Charters, 360-378-2110, of Friday Harbor will pick up parties of two or more from the ferry at Orcas for salmon **fishing**, **sailing charters** (skippered or bareboat), or **whale watching**. The two web sites at the beginning of this chapter list additional fun recreational opportunities for land or sea.

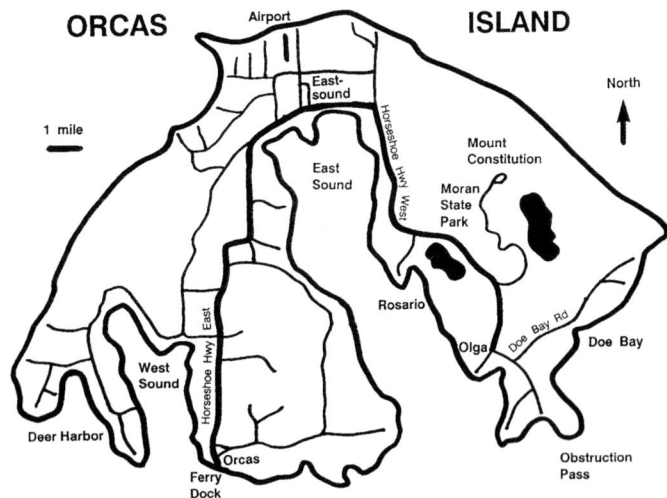

WA - PACK

Packwood

Packwood Airport

Town	Packwood	**ID**	55S
Coord	N46-36.24, W121-40.67		
Elev	1053 feet		
Runway	1 - 19, 2262 x 38' asph.		
Freq	CTAF-122.9, Lights-5x		
Charts	Seattle sectional		

CAUTION: *Airport information not for navigational use. Trees at either end of runway. Wildlife. Closed in winter.*

> *Packwood is a small mountain community 16 miles south of Mt. Rainier that caters to tourists and flea marketeers. The airport is one block from town. Camp at the strip or stay at one of several inexpensive motels. $20-120*
> *360-494-2311, www.destinationpackwood.com*

Airport Description and Services

The Packwood airstrip is literally one block from the center of town. Although the runway is paved and seemingly of adequate length for most small aircraft, it has trees at both ends. Proper landings require a carefully executed steep approach. When departing, watch density altitude. Mountains are not a factor except at night or in poor weather.

The asphalt runway is smooth and perfectly flat. Although the airport has a small beacon and pilot-controlled lighting, night operations are not recommended for visitors. The airport is closed in winter.

After landing, taxi to the grass and create your own tiedowns. The center of town is one block from the airport – just walk past campground and Packwood Hotel. You can camp at the airport, but the only amenity is a pit toilet. Airport information: 360-497-7044 (not at airport).

© 2001 by ALTA Research

Town of Packwood

After several nights in a tent, I like to take to the air, fly to a nondescript town, and discretely chow down on a greasy breakfast. Packwood fits the preferred profile: near backcountry strips, small town, interesting location, and airport within walking distance. This small community has a few well-off retirees, but the majority support the timber industry or cater to tourists.

Normally, Packwood is a quiet place to visit, a hamburger and pizza kind of town – no gourmet restaurants within 20 miles. Just one block from the runway, Hotel Packwood offers the least expensive lodging. This old hotel, built in 1912, is for those who like historic, rustic facilities. The Peters Inn serves a good breakfast. Phone numbers and other options follow.

Packwood Hotel, built in 1912, one block from the airport

Cowletz Lodge	**CJT**	$51-74	888-305-2185
Hotel Packwood	**HJT**	$24-42	360-494-5431
Inn of Packwood	**CJS**	$54-88	360-494-5500
Mountain View Lodge	**JP**	$41-93	360-494-5555
Peters Inn Motel	**PR**	$47-66	360-494-4000

Features
B Full breakfast
C Continental
H Historic
J Jacuzzi
P Pets OK
R Restaurant
S Swim pool
T Transportation

When I arrived, virtually every nook and cranny was filled with a flea market booth. I had stumbled on Packwood's Labor Day Flea Market extravaganza. The swarm of people and vehicles were way out of character for this small village. I learned that the event is a tradition that has grown in popularity over the years. The merchants hawk most anything: books, antiques, junk, leather goods, tarp systems, cattle skulls, food . . . Fragrances of deep fried delicacies tempt the nostrils – elephant ears, French fries, onion rings. The sign reads "Real Hamburgers." I wonder, "the real $100 hamburger of pilot lore?"

Flea market

Pasayten Wilderness

WA - PASA

Lost River Airport

Town	Mazama **ID** WA36
Coord	N48-38.98, W120-30.12
Elev	2415 feet
Runway	11rt - 29, 3150x85' dirt/grvl
Freq	CTAF-122.9
Chart	Seattle sectional

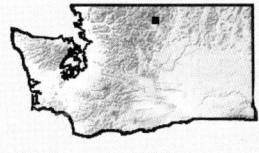

CAUTION: Airport information not for navigational use. Towering mountains nearby. Trees surrounding runway. Deer and vehicles. Watch density altitude.

> *The Pasayten Wilderness offers clear streams, clean air, and challenging mountains to cross. Hiking trails are minutes from the airstrip.*
> *$0-60   Lodge: 509-996-2537*

Airport Description

The airport is situated in a valley between high mountains. The northwest end is located rather close to the side of a mountain. When I landed, runway surface-gravel was large – up to two inches in diameter. Though landing is rough due to the size of the rock, I observed no ruts. Windsock and tiedown area are at the west end of the airport.

When arriving, you may have difficulty closing a flight plan by radio at altitudes of less than 10,000 feet. Given neutral wind conditions, departure to the southeast is recommended; trees are cleared for some distance in this direction.

The airport is neither attended nor aggressively maintained. If you require paved runway, airport services, or rental car, consider landing at nearby Winthrop airport.

Airport Services, Transportation, and Lodging

Phone: owner, 800-552-2999 (not at airport). A car dealership will answer the phone; ask for Arnie Nelson. The Lost River Resort phone is 509-996-2537, and often not responsive.

No fuel, tiedown ropes, nor services. Closed in winter.

© 2001 by ALTA Research

Pasayten Wilderness **WA - PASA**

Transportation is not available and not required for the hike described below. The trailhead is within walking distance. The Lost River Resort has cabins with kitchens for $50-up, but no food of any kind. To get to the resort, walk to the highway and turn right.

Overview

The Lost River airstrip is located on the southern edge of the Pasayten Wilderness. This is a beautiful **backpacking** area, with challenging but well-maintained trails.

The best area topo maps are available from Green Trails. They can be found at Washington sports stores, or may be ordered from Green Trails at 800 762-6277. The Pistol Pass / Shellrock Pass Loop (described below) requires maps *18*, *19*, *50*, and *51*. Trail maps are also available from the ranger stations, as

Aerial view of the Pasayten Wilderness

mentioned earlier. The Early Winters Ranger Station is the closest. It is located several miles southeast of the landing strip and is open for several months during the summer.

Hiking and camping permits are no longer required. For the latest information, contact the USFS at 509-996-4000. For a $5 map, call 503-872-2750. A full compliment of backpacking gear is recommended for treks into the Pasayten Wilderness.

Pistol Pass / Shellrock Pass Loop

This loop is a 42-mile, three day hike through the Pasayten Wilderness Area. This loop is a good match for backpackers who love mountain beauty and are willing to work for it. The trail gains up to 4500 feet in one day. In spite of the challenging topography, you will likely meet other hikers on these trails.

From the airport, walk ten minutes west on the main road. Shortly after crossing the river, you pass a cluster of old log cabins on the right. As the pavement turns to gravel, look to the right for a survey marker tree on which is posted a small yellow sign that reads *Bearing Tree*. Follow a shortcut path near the sign and bear right when it joins the main trail.

Follow the Monument Creek Trail north. After about one hour, the trail passes close enough to the river for access to non-designated riverside campsites. After one and one-half hours from the trailhead, you cross the Eureka Creek bridge. Several conventional campsites are available just after the bridge. This is the last campsite with water for quite some time. There were no mosquitoes when I visited.

Before continuing, load up with two quarts of water per person. What lies ahead is a 4300-foot climb, with water unavailable until after Pistol Pass. The climb to the pass takes about six hours. Campsites are available, but not with water.

© 2001 by ALTA Research

Pasayten Wilderness

WA - PASA

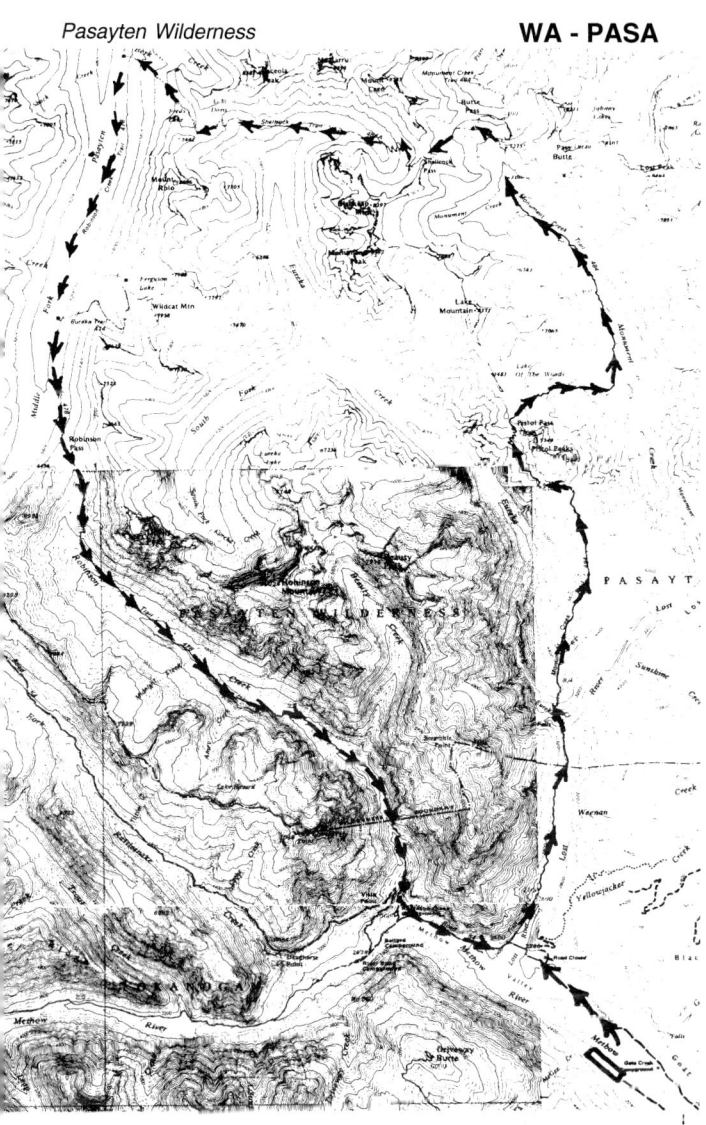

About 20 minutes past Pistol Pass, look for a sign that says *1/4 Mile To Lake Of The Woods*. This is the first source of water – a scenic campsite.

From Lake of the Woods, traveling time to Monument Creek crossing is 1.5 hours, mostly downhill. Past Monument Creek, water is available at frequent intervals. Continuing on, total travel time from the Lake of the Woods to the intersection of Shellrock Trail is 3.5 hours. After crossing Monument Creek, it's all uphill until the intersection. Follow Shellrock Trail another hour uphill until Shellrock Pass. Elevation gain from Monument Creek to Shellrock Pass is about 2800 feet – 4.5 hours since leaving the Lake of the Woods.

The view from Shellrock Pass is panoramic. Continue west on Shellrock Trail toward Freds Lake. Drop down into a valley with views of rugged, towering mountains. Climb up 1200 feet past Lake Doris, over a pass, and down to Freds Lake. Time from Shellrock Pass to Freds Lake is about 3.5 hours. There are suitable campsites at the west end of Freds Lake.

After Freds Lake, walk downhill forty-five minutes to the intersection with Robinson Creek Trail. Follow this trail south. You will find campsites after about fifty minutes of walking. Compared to other segments of the loop, you are entering the least interesting interval. While walking nearly three hours on the stretch between the intersection and Robinson Pass, you experience an elevation gain of 1100 feet. After Robinson Pass, you're on the final downhill leg.

Expect to log about 3.5 hours on the segment between Robinson Pass and the Robinson Campground. Several reasonably attractive campsites are available along the way. Robinson Campground is located on a road that leads to the airport. Follow the road south for 50 minutes until you reach the airport.

- RW

© 2001 by ALTA Research

WA - RANG

Ranger Creek

Ranger Creek Airport

Town	Ranger Creek **ID** 21W
Coord	N-47-00.78, W-121-32.07
Elev	2650 feet
Runway	15 - 33, 2680 x 29' asphalt
Freq	CTAF-122.9
Charts	Seattle sectional (new)

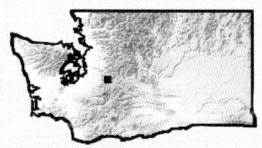

> ***CAUTION:*** *Surrounded by mountains. Changeable weather. Turbulence. Not recommended in winter. Night operations not possible.*

> *Ranger Creek, a mountain strip within easy striking distance of Seattle. An excellent paved strip tucked in a tight valley. 13 miles NE of Mt. Rainier at the corner of the national park. Plenty of campsites within short walking distance of a glacial river. $0*

Ranger Creek Airport

Squeezed between towering mountain sides, the runway lies parallel to and west of the White River. Besides being a strategically-placed safety strip, it is an excellent site for mountain training. Turbulence and claustrophobic maneuvering space make the experience real, while the long, unobstructed, perfectly flat runway makes for a safe touchdown platform.

The airport is located at the northeast corner of the Mt. Rainier National Park boundary. Unless the discrepancy has been corrected, topo maps show the location one mile north of the official coordinates supplied by Washington Aeronautics.

While Ranger Creek weather is statistically better than Seattle or coast, it can shift from scattered to overcast in minutes. Mount Rainier has significant influence over the weather, both good and bad. Plan ahead: Don't visit this site on a cloudy day without a sleeping bag. If the clouds close in, resist temptation to make a mad dash for a "sucker hole." Wait for a stable patch of blue.

The primary factor for runway selection is wind. The runway is flat, open at both ends, and provides safe go-arounds in either direction. Either way, you need to make a tight base turn several miles or more from the runway. After landing, pick a convenient tiedown area and pound your own.

© 2001 by ALTA Research

Ranger Creek **WA - RANG**

Washington Aeronautics: 206-764-4131. At Ranger Creek, there are no lights, no services, no fee, no food or restaurants, no lodging – just campsites, glacial river, mountains, and fresh air.

Ranger Creek Campground

The reopening of Ranger Creek airstrip has raised a cheer and sigh of relief from mountain-loving pilots in the area. Originally, the strip was used during the Korean War as a communications outpost for training. It remained in use until 1991 when the U.S. Forest Service refused to reissue its use permit. With the help of the Washington Pilots Association, Washington Aeronautics spent four frustrating years shuffling paper and attending various environmental impact hearings.

Although Ranger Creek is located in a beautiful mountain valley, you share the scenery with car and RV campers. Even so, there are plenty of nice campsites under the dense canopy of evergreens between runway and White River. The grounds are a camp-maker's dream: sun and wind protection, flat, brush-free, a vast bed of pine needles and moss.

In spite of the scenery, my first impression was, "Too many people." After chatting with a dozen or so, I amended it to, "Too many nice people." They were all "regulars," families who return to this spot year after year. All were glad the airstrip was reopened. "Adds some excitement," they say.

I was offered pop and snacks, and even an invitation to dinner. Besides the potential for camaraderie, the old timers have a wealth of information. They can describe hiking trails, point you to the trailheads, and spin local lore. One old timer swears there was a Cold War era missile site just north of the field – might be worth investigating.

Hiking options range from easy to strenuous. For easy, you can stroll the White River. Though its glacial silt makes for marginal fishing, no license is required south in the park. A more aggressive trail takes you from the airstrip, west to Fawn Ridge. Serious hikers may climb the 12-mile trail northwest to Sun Top Lookout at 5269 feet. The venerable Pacific Crest Trail passes the airstrip only six miles east, but actual access is over double this distance by trail.

Mount Rainier's alpine Sunshine Visitor's Center is located at 6400 feet, only 15 miles by road. If you have the time, why not swap a plane ride for a car ride to this scenic vista?

WA - ROCH

Roche Harbor

Roche Harbor Airport

Town	Roche Harbor **ID** 9S1
Coord	N-48-36.7, W-123-08.2
Elev	60 feet, E end high, 0.7%
Runway	6-24, 4180 x 30' oiled grvl
Freq	CTAF-128.25
Charts	Seattle sectional

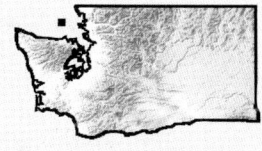

CAUTION: *Airport information not for navigational use. Dip in middle of runway; east end high. Deer.*

> *Roche Harbor is a picture postcard environment – an old-fashioned resort on the ocean with gourmet dining, evening entertainment, historic walks, boat rentals, swimming, tennis, and volleyball. Nearby: beach combing, tide pools, fishing, snorkeling, and scuba diving. $75-400 Resort: 800-451-8910, www.rocheharbor.com*

Airport Description

Roche Harbor Resort requests pilots to phone before landing. The air field is located at the north end of San Juan Island. The long strip radiates from the western harbor resort into forested terrain to the east. Because the east end is 60 feet higher, land on Runway 6 and depart on 24 (unless strong winds suggest otherwise). The owners generally keep runway lights burning until 23:00, after which you can request lights by phone.

Parking areas are located at the west end and 1000 feet west of the east end. Tiedowns are not provided at the west end. Due to ground-slope in the tiedown area, take care to apply brakes and chocks. Deposit $5.00 landing fee in the box at the west end. Ask for reimbursement if you stay overnight at the hotel.

Airport Services, Transportation, and Lodging

Airport phone: 800-451-8910. Field unattended. No fuel. Call the resort before landing. Advise them of your plans and ask for (free) transportation upon arrival. If your luggage is light, the walk to Roche Harbor is short and pleasant. From the west end of the runway, follow the roads five minutes northwest to the harbor and hotel.

© 2001 by ALTA Research

Roche Harbor **WA - ROCH**

Roche Harbor Resort provides a wide selection of lodging, ranging in price from $75 to $323 for two, depending upon amenities and season. The historic Hotel De Haro, which served as Hudson's Bay trading post in the 1850s, now provides rooms with antique decor and showers down the hall for $75-96.

Call several weeks before your visit to secure a room with a view of the harbor. The weekends and holidays of June and July are often fully booked in advance.

Official campsites are not available in the area. However, seasoned visitors discretely camp at high points in the nearby quarry. An employee of the resort confessed that while parents indulge in comforts of the lodge, some allow their children the adventure of tenting at the quarry.

The quarry has natural campsites tiered on the edges of its walls. Although not a safe choice for small children or sleep walkers, these vantage points offer magnificent views of the surrounding islands. At dusk, the ledges provide premier seating for nature's evening pageant of colors.

To get to the quarry, walk southeast from the air strip on Roche Harbor Road. This walk takes you up a rise, past a resort entrance on the right and past a large green propane tank on the left. After a total of ten minutes, you will be at the top of a second rise. Look for a rectangular concrete water tank with green roof on the right. Follow a dirt path left and up hill into the woods. Continue walking for several minutes as the path gains elevation, circling the outside of the quarry clockwise. As it climbs, you will be able to branch to campsites that sit on ledges overlooking the quarry. Continue to the top for campsites with the most spectacular views.

© 2001 by **ALTA** Research

Roche Harbor Resort

Roche Harbor was originally built around thirteen quarries containing some of the richest lime deposits in the west. In the early 1880s, John S. McMillin created the Roche Harbor Lime and Cement Company, the largest limeworks west of the Mississippi. Production continued until 1956 when the Tarte family purchased the property to create a resort. For over thirty years, the Tartes worked to bring their dream to fruition.

The Tarte family has been successful at preserving the history and charm of Roche Harbor. Buildings and rooms maintain their old-time flavor. Though facilities lack the amenities of a first rate modern resort, they have an old fashioned charm with an immeasurable value of its own. Graced with subtle landscaping, award-winning flower gardens, and a marina with boats of all shapes and sizes, Roche Harbor is a picture postcard environment.

Hotel de Haro marina

Roche Harbor **WA - ROCH**

    Tradition and lore are a major part of the Roche Harbor experience. Local fact and fiction are fascinating. Ask the hotel attendant for the pamphlet entitled *A Walking Tour of Historic Roche Harbor*. Take a couple of hours to **walk the site** from quarries to ghostly mausoleum. Among the practiced rituals are the **flag ceremony** at sundown and the recorded church chimes from the chapel. I cannot find words to describe the feeling of walking into the harbor for the first time, hearing echoes of church chimes playing *My Favorite Things*. It's almost eery.

    The harbor can accommodate over 500 boats. During festival seasons, moorings fill to capacity. You can **rent a boat** and motor to picturesque islands within one or two miles. On-site **tennis**, **volleyball**, **swings**, and **swimming** are available. Changing-rooms at the pool are Spartan, with no place to lock valuables. If you do not stay at the resort, you can have access to the resort's pool and tennis court for a fee.

Hotel de Haro

The marina has a number of facilities that can be useful to travelers: laundry, showers, general store, and snack bars. **Inexpensive meals** are available at the pier and swimming pool. The hotel's restaurant offers **gourmet dining** with views of the harbor. Meals cost $9-12 for breakfast, $9-12 for lunch, and $18-35 for dinner. **Evening music entertainment** is provided in the lounge below the restaurant on weekends in summer.

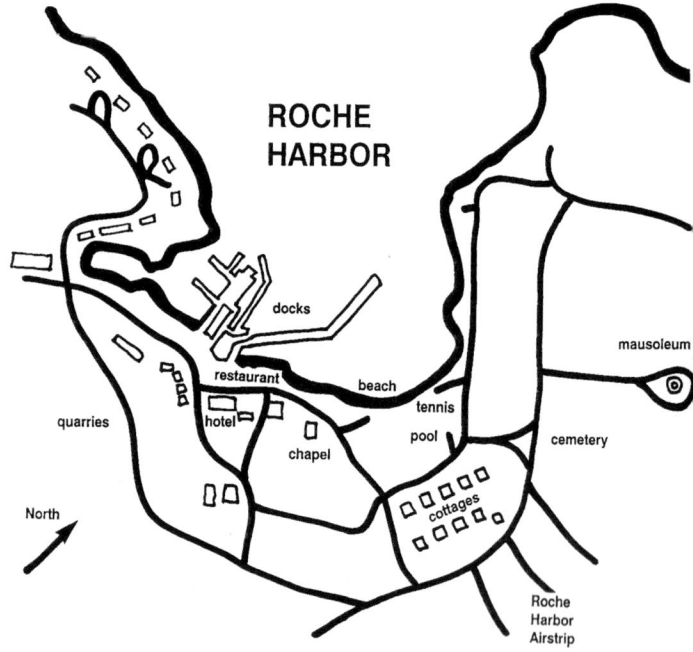

Roche Harbor **WA - ROCH**

Surrounding Area

Rent a small motorboat and you gain access to the scenic complex of local islands west of the harbor.

Posey Island is perhaps the country's smallest **marine state park**: one acre, one campsite, one toilet, and picnic tables. Activities around the islands include **beach combing**, pondering **tide pools**, **snorkeling**, **scuba diving**, and enjoying rich sunsets. Posey Island lies on the far side of Pearl Island, less than one mile northwest of Roche Harbor.

The historic **British Camp** is also within boating distance. Sail south through Mosquito Pass to Garrison Bay and dock at the southern edge of Bell Point, about two miles total. You can also reach the Camp on foot. Follow Roche Harbor Road and West Valley Road two miles south to British Camp. Sites of interest include a walk around Bell Point, restored buildings, relics, and a slide show. Local history is a microcosm not dignified by high school text books. You will learn how an ambiguously-worded treaty and an irreverent porker triggered the Pig War in 1859.

While at British Camp, don't miss the **short hike** to the peak of Mt. Young, the most spectacular viewpoint on San Juan Island. The hike to the peak is less than one mile and elevation gain is 590 feet. The trail begins at the picnic area on the northwest corner of the parking lot. From the trailhead, walk five minutes to cross West Valley Road and another five minutes to the British Military Cemetery. A short trail at the northeast corner of the cemetery joins the trail up Mt. Young. The trail climbs the north face, switchbacks south, and zigzags to the summit. The views are a fabulous mosaic of islands and water, with a backdrop of snowy peaks on the mainland.

- RW

© 2001 by **ALTA** Research

View from Hotel de Haro balcony

WA - TIET

Tieton

Tieton State Airport

Town	Rimrock **ID** WA49
Coord	N46-28.27, W121.07.42
Elev	2961 feet, N end high
Runway	2 - 20, 2471 x 140' turf
Freq	CTAF-122.9
Charts	Seattle sectional

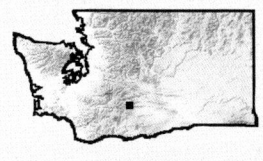

CAUTION: Airport information not for navigation. Rwy 20 requires nonstandard approach, due to mountain obstruction. Expect turbulence. Soft when wet. Wildlife. Closed in winter. No night operations.

> *Tieton sits at the edge of Rimrock Lake, 28 miles southeast of Mt. Rainier. Dramatic approach and challenging touch-down. Scenic setting and a variety of campsites. No food or lodging. $0*

Tieton State Airport

Tieton (Ty-a-ton) airstrip is located at the east end of Rimrock Lake, visible as you cross the mountain ridges that surround the lake. Though Goose Egg Mountain appears to block the northeast end of the runway, experienced mountain pilots will quickly spot an unorthodox approach.

As the strip came into view, my first reaction was, "Hey, just fly in over the lake and land uphill on Runway 2. Then the brain clicked in. "Reed, you had better think this over. The wind is from the southwest, and the south exit over the lake provides a safe go-around..." OK, I opt for the unorthodox approach to Runway 20. Lots of flaps, steep turn, low approach, air bumps, touchdown. Romeo rolls smartly to a stop at midfield.

The runway's north end goes up hill, the middle is flat, and the south end drops off to the lake. The surface is turf, with grass recently cut to four inches. Sections of the runway have two-inch ruts, which may explain why the airport is closed October through June. Droppings of horse and elk indicate that large animals frequent the runway.

Airport Information: Washington Aeronautics, 206-764-4131 (not at airport). Other than bare-bones camping necessities like pit toilets, there are no services at or near the airport. Operation in poor weather or at night is not recommended.

© 2001 by ALTA Research

Tieton **WA - TIET**

Rimrock Lake

After landing, I hear that sweet sound – a low-flying Super Cub passes over head. He validates my choice by selecting Runway 20. The southerly wind balks at being deflected east by Goose Egg Mountain, giving the Cub a bumpy final. The Cub rolls by, filled to the brim Alaska style.

The pilot is a Tieton regular. He and his 160 HP Super Cub had spent 15 years flying in Alaska. While unloading provisions, he commented that Tieton is "considered to be a hazardous strip." More accurately, without proper skill, judgment, and equipment, the pilot would be considered hazardous at this strip.

Washington Aeronautics has kept the two large windsocks and runway markers in tip-top shape. Park anywhere at the side of the runway. For liability reasons, Washington Aeronautics does not install tie-downs at emergency strips – a novel concept.

Campsites of all styles are located around the strip: lakeside, open, or heavily treed. My preference is the shady campsite at

the western edge of the strip. It includes a private outhouse, which is probably intended for pilots. When I visited, a family had staked their claim with two tents, campfire, and living room couch. They said I could join them if I liked.

Although there is no lakeside snack bar, you may be able to pull a few fish out of Rimrock lake. There is an adequate supply of wood for cooking your own over a camp fire.

The lake and surrounding camp attract outdoor enthusiasts with a plethora of motorized toys: speed boats, jet skis, dirt bikes, and ATVs. Pilots, with the biggest and best of all, should feel right at home. On the other hand, if you like peace and quiet, you'll have to take a hike. Trails radiate in all directions. One regular camper pointed out the options, "There's one there, another there, and over there a blocked road that goes up into the mountains."

During summer, the weather is generally pleasant. However, conditions can get windy or hot, hazardous for flying. Due to local weather, the mountains to the west can be obscured in clouds.

Wyoming

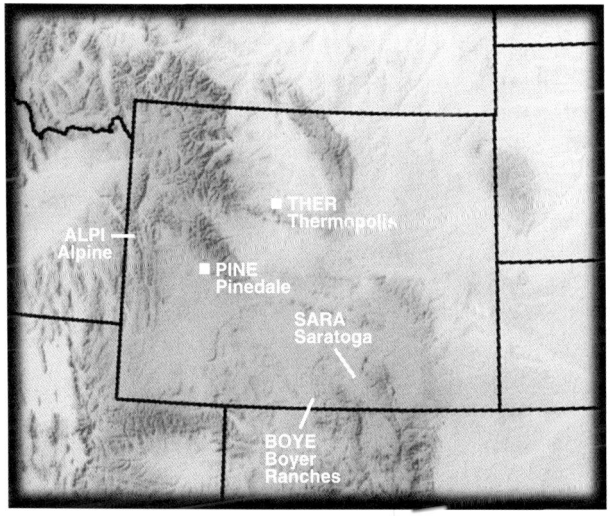

Wyoming Aeronautics

307-777-3952

Wyoming

© 2001 by ALTA Research

WY - ALPI

Alpine

Alpine Airport

Town	Alpine	**ID**	46U
Coord	N43-11.08, W111-02.55		
Elev	5634 feet		
Runway	13 - 31, 5625 x 35' asphalt		
Freq	CTAF-122.8		
Chart	Salt Lake City, L7, L8, L9		

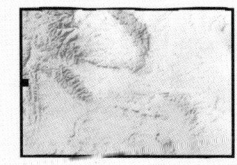

CAUTION: Airport information not for navigational use. Airport closed for night operations. Last 50 feet of SE end slopes down 5 feet.

Alpine is an eye-blink junction that offers hiking, rafting, and wild times at a local bar. $0 - $200. Info: www.starvalleywy.com

Airport Description

The airport is located near the southeast end of Palisades Reservoir. It is clearly visible and free from local obstructions except for mountains two miles to the northeast. Runway length and surface make for a safe landing. The airport is closed at night. The tiedown area is not surfaced. Bring your own tiedown hardware. There is no tiedown fee.

Airport Services, Transportation, and Lodging

Airport information phone: 307-654-7660. No services or telephone are located at the strip. Food, lodging, camping, and entertainment are within walking distance. Rental cars are unavailable at Alpine. There is no conventional taxi service, but residents will drive you to a trailhead for a fee.

The Nordic Inn and Royal Resort provide the closest source of food and lodging. The walk from the airport to the inns requires 15 minutes. Follow the road northeast for 10 minutes. Turn right on Highway 26 and walk another 5 minutes. You can't miss. The Three Rivers Motel is located a tad farther, and it seems to be run in a friendly and down-to-earth manner.

The Best Western Flying Saddle Lodge is also about one mile from the airport. Open May through October, they have a pool, tennis court, and gourmet dining. Reservation: 800-528-1234. The Box Y Lodge and Guest Ranch, 30 miles from town, includes all meals and has equestrian activities.

Features
- **B** Full breakfast
- **C** Continental
- **H** Historic
- **J** Jacuzzi
- **P** Pets OK
- **R** Restaurant
- **S** Swim pool
- **T** Transportation

© 2001 by ALTA Research

Alpine **WY - ALPI**

Alpine Inn (cabins)	JP	$53-89	307-654-7644
B.W. Flying Saddle Lodge	JPRST	$74-89	307-654-7561
Box Y Lodge & G. Ranch	BHJRT	$200	307-654-7564
Nordic Inn	PRT	$80-89	307-654-7556
Royal Resort	JPRT	$58-89	800-343-6755
Three Rivers Motel	PT	$50-50	307-654-7551

Lakeside camping areas are located one mile west of the airport and two miles northwest off Highway 26. I simply camped under the wing of the plane.

Food, Entertainment, and Recreation

The town has grown and settled since my last visit. I fondly remember wild weekends at Jeep's Bar. The food wasn't special, but Jeep's drew big-name country western entertainers from all over the west. Now called the **Bull Moose Saloon**, 307-654-7594, they hire local or road bands that play on weekends. To get there, you need to walk 20 minutes from the Nordic Inn. Follow Highway 26 south. Highway 26 turns into Highway 89, crosses the Snake River, and you're there.

Other culinary options include tasty pizza at Gunner's Pizza, Red Baron Restaurant, Cringle's, and the Flying Saddle Lodge.

Various Jackson outfitters provide half-day **whitewater raft trips** through the "Grand Canyon" of the upper Snake River. If you have never experienced river rafting, this is a quick and inexpensive way to get a good taste of the thrill. Arrange the trip and pickup through your inn, or call the provider direct.

Barker-Ewing River Trips	*rafting*	800-448-4202
Deadman Creek Outfitters	*hunting*	307-654-7528
Jackson Hole Whitewater	*rafting*	800-700-RAFT
Mad River Boat Trips	*rafting*	800-458-7238
Sheet Mountain Outfitters	*hunting*	307-654-7564

© 2001 by **ALTA** Research

Short Hike

The following hike requires 60 to 90 minutes to reach one of two end points. Elevation gain is 750-900 feet, depending upon how far you go. The destination is a mountain side campsite with a view of Palisades Reservoir. Bring your own water.

Follow the airport road northeast for 10 minutes. Turn left and walk another 10 minutes northwest on Highway 26. Six hundred feet past the Alpine Point of Entry checkpoint building, turn right on a dirt road that heads northeast toward a building with *Alpine Storage* painted on its side. Follow this road past a couple of cabins and into the forest.

The road becomes an old 4x4 trail. In June the ground was covered with yellow and purple flowers. The forest is lightly populated with pine trees and aspens, rendering about 50% shade on the trail. The mountains are visible above the trees.

You reach the first vista point one hour after leaving the airport. This site is suitable for a tent, but is not close to water. The view from this elevation of 6375 is airport and reservoir.

Alpine

You may push onward to a higher vantage point. Continue walking northwest. Bear left at each intersection. The trail ends at elevation 6525 with a better view of the lake and town. The campsite has a good view, but again lacks water.

Spring Basin Hike

This hike takes you into the neighboring mountains. The uphill leg to an alpine meadow at 7350 feet requires about 2.5 hours. The return can be done in less than 1.5 hours. Optionally, the hike can be extended to a several-day loop. The lower elevations of this hike serve as a cross-country area in winter.

Begin as explained in the "Short Hike." Continue northwest on Highway 26 for 30 minutes after leaving the airport road. (You can save some time by taking a short cut from the runway, north to Highway 26.) You pass BG's Alpine Merc on the left, a source of food and supplies. Turn right on to a dirt road 600 feet past the store. A nearby sign reads *Targage National Forest Organization Camp*, and *Alpine 4H*. Follow the dirt road to the 4H Camp. Turn right and walk east through the camp.

Continue northeast through open, flowered meadows and cross a mountain creek. You pass several good campsites along the 4x4 trail. While crossing the bridge over a creek, a sign reads *S. Fork Indian Creek*. Take the left branch where the trail splits. The trail winds through meadows and stands of trees as it finds its way up the valley. At 6400 feet, the 4x4 trail turns into a foot trail that begins the mountain ascent. The trail is marked with double ax marks on trees.

The steep ascent ends at an alpine meadow at 7350 feet. Water is available, as well as a number of flat, protected tent-sites. Old metal watering troughs mark the entrance to the meadow. From this point, you can head back from whence you came or continue in a several-day loop. If you opt to continue, be sure to have a topo map and some good local advice.

© 2001 by ALTA Research

Boyer YL Ranch

WY - BOYE

Savery Creek Thoroughbred Ranch

Boyer 'International' Airport

Town	Savery **ID** none
Coord	N-41-07, W-107-23 est.
Elev	7000' est. N high, mid. low
Runway	6-24, 4500'x30 est. dirt/turf
Freq	CTAF-122.9
Charts	Cheyenne sect. (not shown)

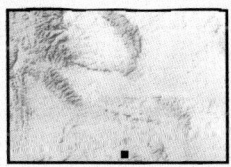

CAUTION: Airport information not for navigational use. Runway surface not routinely maintained. Check for livestock, antelope, and deer. Watch density altitude.

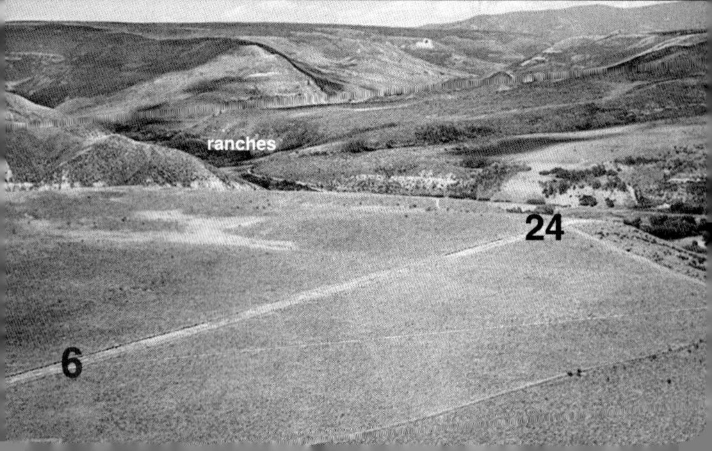

> *Twin ranches run by branches of the same family. Boyer YL offers world-class mountain bike recreation and photography workshops; Savery specializes in English and Western trail riding, dressage, jumping, and cross country. Both offer training on the trail.*
> *Boyer YL: 307-383-7777, $300-500*
> *Savery: 307-383-7840, $250-400*

Airport Description

The airstrip is located 10 miles east-northeast of Dixon Airport, 9U4. Look for the runway on a plateau, several hundred feet above and just west of the green river valley below. There are no buildings or windsock at the strip.

If you have concerns about a safety risk, land on the generous runway at Dixon instead. The Dixon runway is paved and has pilot-controlled lights on 122.9. Either way, you receive free pickup. By prior arrangement, buzz the ranch buildings by the river, one mile east-northeast of the airstrip.

The Boyer tiedown corral has one set of rope-less tiedown stakes near the "Boyer International" sign. Bring your own ropes and stakes, just in case. With the corral open, aircraft are not protected against curious livestock. Pilots with fabric airplanes would be wise to secure the corral with a couple hundred feet of rope.

Airport Services

Airport phones (at ranches below): Boyer YL, 307-383-7777; Savery Creek, 307-383-7840. There are no phones or services at the strip. If you call prior to departure and buzz on arrival, someone from the ranch will meet you.

Boyer YL & Savery Creek Ranches **WY - BOYE**

Overview

Before the U.S. and its Homestead Act, this area was a sanctuary and hunting grounds of the Ute Indians. White man arrived, and decided the land was more fit for sheep and other livestock. The Boyer site was first homesteaded by a German couple in the 1890s.

John Boyer purchased the ranch in 1905. He expanded the facilities, married, and opened southern Wyoming's first guest ranch in the mid 1920s. At that time, guests rode horses a day and a half through the high country over the Continental Divide to get to the ranch.

After years of dormancy, the ranch has been divided between Boyer descendants. The two resulting ranches were reopened to the public in 1996. They are being developed in unique ways by various family members with diverse skills. The owners realize they have inherited the family treasure, and they are working hard to give it new life.

You will likely meet up to three generations of family members. Be prepared for frank discussions on a variety of subjects. Senior members are as sharp as a tack, and like most folks from Wyoming, they will not mince words. Don't be surprised if they test your bent in a political debate soon after you meet.

The ranch land joins a vast expanse of National Forest land to the east and the Red Desert to the west. This enables unlimited **horseback riding**, **biking**, and **hiking** in nearly all directions. Birds and other wildlife are abundant. Bands of wild horses are still at large in the Red Desert.

Neighboring towns offer a taste of **wild west history**. This is Wild Bunch country. You can still visit **Butch Cassidy's cabin** in Baggs, just a few miles from the ranch. Check your sectional and this book for other interesting excursions within easy striking distance by air.

© 2001 by ALTA Research

Frankly, I think you'll find plenty to do within riding distance of the ranch. Because the two ranches work together, you experience variety that far exceeds the norm. Each day you can try something different: **horseback trail riding, jumping, biking, photography, tennis, badminton, croquet, dipping in the creek**, and don't forget to plan some TLC for those tired muscles!

Boyer YL Ranch

The Boyer YL Ranch is currently operated by Jock, Winston, and Elisa Boyer, descendents of the original John Boyer. They have achieved worldwide recognition.

Winston is an accomplished professional photographer who has authored a stunning book of scenic photos. He lends his expertise to **photography workshops** at the ranch. Ask him to show you examples of his color commercial and art photographs.

Jock is a world-class bicycle racer. He was the first American to race in the Tour de France, and has a long list of wins to his

Boyer YL ranch house

credit. He has channeled his expertise into an import/export business that specializes in high-end bicycle components.

Take a beautiful ranch in the foothills, add biking expertise and a couple of guys with entrepreneurial spirit, and what do you get? A **mountain biking ranch**, of course. What a great idea! The **bike rentals** and mechanics are handled on site by Sore Saddle Cyclery from Steamboat Springs. Two **pro-riders** will be on hand to shame you into riding your best. A **masseur** and **gourmet cook** will be awaiting your return from the trail.

Not every family member wants to peddle for hours in the mountains. Other options include **hikes**, **swimming hole**, **tennis**, **croquet**, **photography**, **fly fishing** and **trail rides** on the Savery Creek Thoroughbred Ranch horses for a fee.

The Boyer YL buildings were constructed years ago of hand-hewn logs. Yet, the facilities have been sufficiently modernized to meet the needs of all but the most finicky visitors. In the ranch house, you sit on solid antique furniture before a huge stone fire-

Boyer YL ranch house

place. The original log walls are hung with artifacts and animal trophies. A gigantic moose head above the fire place seems to preside – if only he could talk!

The little cabins, circa 1920s, are charming and authentic. Each has a stone fireplace, as does an additional sleeping room in the ranch house. Meals are prepared and served at the lodge by a California chef. I did not have an opportunity to enjoy a Boyer YL meal, but the owners are aiming toward gourmet.

The Boyer family is apparently still developing a direction for the ranch. Their rate structure was not fully defined at the time of this update. I expect they will ask $300-400 per couple per day, meals included, activities extra. The minimum stay is typically three days. As a courtesy, transient pilots may stay one night, subject to availability. When you call for information, be sure to verify recreational options and specific costs.

Boyer YL Ranch guest cabins

Savery Creek Thoroughbred Ranch

The Savery Creek Thoroughbred Ranch will be of particular interest to visitors who love to ride **horseback**, both **English** and **Western**. The ranch's well-trained, responsive horses will be especially appreciated by experienced riders. English style and jumping are specialties of the ranch. Guests may receive **training** in **general riding**, **dressage**, or **jumping** while on the trail.

Joyce Saer is the dynamo behind the Savery Creek operation. Joyce, daughter of John Boyer, has worked horses all her life. She has lived in various parts of the U.S., Spain, England, and Ireland, and brings these equestrian cultures together at her ranch.

Joyce enjoys exposing people of all ages to the satisfaction of being a good rider. Her two full-time wranglers were young women. Both knew the back trails and handled their animals with grace. The young wranglers and Joyce's grand children

Savery Creek Thoroughbred Ranch house

were clearly an inspiration to the children of adult guests. One guest confided that his kids outrode him two-to-one!

Savery Creek Ranch is a working ranch with sheep and some 35 horses. Most are thoroughbreds of competition quality, and many are excellent jumpers. And, if you fall in love with Trigger, you can take him home. Most of the **horses are for sale**, offering an opportunity to try before you buy.

The ranch springs to life before the sun clears the hills. Horses get restless and ease their way across the river toward the barn. The stallion perks up when he hears splashing hoofs. Wranglers scope the scene as they arrive for breakfast and chores. Sunbeams blast over the ridge, causing bright-colored flowers to explode into view – almost like someone cranked the color control to max, I muse. I smell coffee, and head toward the source.

I join a din of morning chatter around the breakfast table, mostly plans for the day punctuated with one-liners and laughter. There's plenty to eat, and the cook-cum-housekeeper prepares Joyce's secret pancake recipe to perfection.

Savery Creek bedroom

Boyer YL & Savery Creek Ranches **WY - BOYE**

Breakfast and lunch are tasty, though not gourmet. Lunch is usually served outside. This leverages the setting and pleasant high-country climate, at the expense of being buzzed by flies. Cookouts and picnics at the Red Desert add a touch of spice. Evening meals are more elegant than the norm. I particularly enjoyed Joyce's succulent lamb, served with wine to candle light on antique china – a gourmet delight.

Joyce consciously maintains a focus on the riding aspect of her business. Real horse people, she says, will be so tired after a day of hard riding, that they'll be glad to flop into a comfortable bed after a good dinner. The trade-off is that accommodations and buildings fall short of five-star perfection. The grounds have a variety of old gear in almost every direction you look; some folks will see old rusty farm implements, others will see historic antiques.

The gate latches tell the story. They are all handmade, inconsistent, and non-ornamental – a chain on a nail, for example. Ah, but they all can be efficiently operated from horseback. This is a

non-pretentious ranch, complete with dust, flies, utilitarian construction, and some of the world's best riding within easy access of a small airplane.

The outside of the lodge looks typical Wyoming – reverse batten board siding with ample shaded porch. I made good use of this resource during the lazy hours. Imagine my surprise when I kicked back, picked up a newspaper, and found yesterday's Wall Street Journal in my hands.

When I called Joyce to get her current prices for this update, she was unwilling to quote prices over the phone. Expect $250-400 for doubles. Carefully verify rates and terms when you call. Considering that you can ride excellent horses all day without paying extra at this ranch, the rates seem reasonable.

You, or more likely your kids, may enjoy sleeping in an authentic **sheep wagon**. These units will provide a memorable, if primitive, experience. Although there is generally a three-day minimum stay at the ranch, pilots who are passing through may want to call ahead to see if a single overnight stay is available.

Wine is provided at dinner, but liquor is not sold on the premises. No smoking inside, and no children too young to ride. Call Joyce for further details regarding refund policy, riding experience, weight limitations, clothing, and safety gear. If you ride, be advised that Wyoming is an at-your-own-risk state.

- *RW*

© 2001 by **ALTA** Research

WY - PINE

Pinedale

Ralph Wenz Field

Town	Pinedale **ID** PNA
Coord	N42-47.73, W109-48.43
Elev	7085 feet
Runway	11 - 29, 7100 x 100' asph.
Freq	CTAF-122.8, Lights-122.8-3x5x7x
Charts	Salt Lake City sect, L8

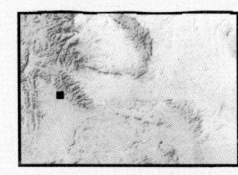

CAUTION: Airport information not for navigational use. Birds, ultralights, helicopters, and deer. Watch density altitude.

11

29

> *Pinedale, home of the Green River Mountain Man Rendezvous, and gateway to the Wind River Mountains recreation area – great hiking, fishing, and hunting. $80-400*
> *307-367-2242, www.pinedalechamber.com*

Airport Description

The airport is 5 miles southeast of town, easy to locate with the POY NDB at 344 KHz. Runway lights and VASI are pilot controlled. The runway is plowed in winter.

Airport Services , Transportation, and Lodging

Airport phone: New Breed Aviation, 307-367-2290. Fuel: 100 and Jet. The bathrooms are always open, and camping is allowed. Hangar and preheat are available for winter visitors.

New Breed has a courtesy car for lunch runs. Pinedale Taxi, 307-367-7668, provides conventional taxi service. Great Outdoor Transportation, 367-2440, specializes in rides to trailheads for $20-70. Allen Dodge, 367-6608, is the source of rental cars in town.

Pinedale has a variety of lodging. The motels and B&Bs are primarily located in town, or at least close to town. The lodges and guest ranches are up to 30 miles from town center. Many of the inns are closed in winter.

The Lakeside Lodge is four miles north of Pinedale and sits on glacier-fed Fremont Lake with a marina and restaurant. They rent motel rooms, primitive cabins, and campsites. The facilities are a tad funky, but accommodations are adequate and the views are very picturesque. The Wind River Mountains across the lake are snow capped, even in summer.

Pinedale **WY - PINE**

Name	Features	Price	Phone
Chambers House B&B	BHPT	$59-126	800-567-2168
DC Bar Guest Ranch	BHJT	$297-389	888-803-7316
Fort William Guest Ranch	BHPT	$55	307-367-4670
Green River Guest Ranch	PT	$70	307-367-2314
Half Moon Lake Resort	CRS	$90-112	307-367-6373
Lakeside Lodge, marina	PRT	$70-138	877-755-5253
Pinedale Best Western	CJPS	$80-123	307-367-6869
Rivera Lodge	PT	$65-107	307-367-2424
Sun Dance Motel	PT	$59-70	800-833-9178
Teton Court Motel	JPT	$49-75	307-367-4317
Wagon Wheel Motel	T	$50-75	307-367-2871

Over 14 campgrounds dot the area. For further information, call the Pinedale Ranger Station at 307-367-4326 or the Chamber of Commerce at 307-367-2242.

Features
- B Full breakfast
- C Continental
- H Historic
- J Jacuzzi
- P Pets OK
- R Restaurant
- S Swim pool
- T Transportation

Mountain Man Rendezvous

Pinedale sits in a valley at the base of the Wind River Mountains, headwaters of the Green River. The town is best known for its mountain man heritage and annual Green River Mountain Man Rendezvous, which fills the town to capacity in mid-July.

It all started because of the beaver hat, a curious whim of fashion that launched thousands of adventurers into the west in the early 1800s. These men, hunters and trappers, were the first white men on the scene. They brought their trapping skills from the east, but were strongly influenced by the Native Americans who taught them how to survive in the challenging western mountain environment.

Between 1825 and 1840, the Rendezvous met a half-dozen times for several weeks during summer. Mountain men streamed

© 2001 by **ALTA** Research

out of the hills, met, traded, drank, fought, and swapped lies around the camp fire. These events were the pinnacle of one of America's most interesting and colorful cultures. It is this special breed of Americans that Pinedale residents have devoted themselves to remembering.

Since 1935, Pinedale has honored the mountain man. In mid-July, the town is flooded by bearded mountain men in authentic leathers. Tepees pop up, horses in the streets, wagons, musket fire, the smell of spent black powder – color and action everywhere.

On the second Sunday in July, the Pinedale community reenacts the days of the 1830s, when trappers, traders, and Indians gathered for the Green River Rendezvous. Their **two-hour pageant**, surrounded by several days of related activities, has been performed by mountain man enthusiasts for over a half-century.

Mountain Man Rendezvous

Pinedale **WY - PINE**

The Sublette County Historical Society has created a beautiful museum to honor the Mountain Man. The Museum of the Mountain Man sits on a hill outside of town on the way to Fremont Lake. Admission is $4, and they are open daily during summer. Call 307-367-4102 or the Chamber for information.

The Wind River Range is arguably one of the best places to backpack in the U.S. The range includes fifteen of Wyoming's sixteen highest mountains, the highest peaking out at 13,804 feet. The Bridger Wilderness has 400,000 acres, 27 active glaciers, 1,300 alpine lakes, and 600 miles of trails. Fish are plentiful and, of course, the skies are clean and clear.

This beautiful natural resource is protected and available for hiking, rock climbing, riding, pack trips, camping, and picnicking. Call the Pinedale Ranger Station for information, 307-367-4326.

A romantic moment at the Lakeside Lodge, Fremont Lake

You can enjoy the scenery by driving a **26-mile loop** to Elkhart Park and Skyway Drive. Begin at the east end of Pine Street. The Museum of the Mountain Man is your first stop. See local history displays and a wonderful collection of guns, Native American artifacts, trade goods, and trapping gear.

Four miles from the museum, you enter the Bridger-Teton National Forest near Fremont Lake. The lake has 22 miles of shoreline and can be seen from scenic overlooks or by visiting Sandy Beach or Lakeside Lodge. You can **swim**, **rent boats**, **eat**, **camp**, or stay at the lodge. At the end of Skyline Drive, you'll arrive at Elkhart Park, a major entrance to the Bridger Wilderness. Here, you can **camp**, **day-hike**, or embark on a **backpacking** trek.

Pinedale is guaranteed to satisfy the **fisherman**. Abundant rivers, lakes, and streams carry all kinds of trout: native cutthroat, rainbow, brook, Mackinaw, golden, and brown. Contact the local Wyoming Game and Fish Department for details, 307-367-4352.

In fall, **big game hunting** dominates the scene: elk, deer, moose, antelope, and bighorn sheep. Nonresident tags are distributed by lottery well before the season. Apply early to Wyoming Game and Fish. The guest ranches listed with lodging provide **fishing**, **hunting**, **horseback riding**, and **pack trips**. See the Chamber web site for a complete list of local outfitters.

Golfers can play nine holes at the Rendezvous Meadows **Golf Course**, 307-367-4252. Or, stroll the shops on Pine Street. McGregors Pub, 367-4443, is a good place for dinner.

In winter, visitors are lured by **snowmobiling**, **cross-country skiing**, and **ice fishing**. You can ski at White Pine Ski Area, 307-367-6606. Snowmobiling in the Green River Lakes area features spectacular winter scenery. The Continental Divide Snowmobile Trail is 600 miles long! You can rent a machine from Big J's Rentals, 307-367-4320. In early February, the International Dogsled Race runs right through town.

- RW

WY - SARA

Saratoga

Shivley Field

Town	Saratoga **ID** SAA
Coord	N41-26.69, W106-49.41
Elev	6978 ft, SW end high, 1.8%
Runway	5 - 23, 8800 x 100' asphalt
Freq	CTAF-122.8,
	Lights-122.8-3x5x7x
Charts	Cheyenne sect, L1, L8, Lo7

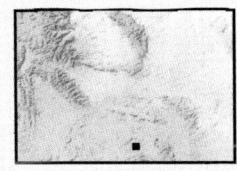

CAUTION: *Airport information not for navigational use. Towers N & S. Wind shear over highway. End of Runway 5 not visible from taxiway; use radio. Antelope. Watch density altitude.*

> *Saratoga is nested in "The Good Times Valley" – a sleepy western oasis with hot springs. A trout stream runs right through town. Stay two hours for a soak in the springs, several days at a resort, or a week on a dude ranch. $70-250 CofC: 307-326-8855, www.trib.com/SPVCC.*

Airport Description

Saratoga's jet-capable runway is visible for miles. The airport is located a quarter mile southwest of town. Usage and service are geared toward business-jet clientele who visit a local, exclusive resort. Don't let this extra-long runway fool you; high summer heat, high elevation, and inclined surface can make density altitude a critical factor.

Airport Services, Transportation, and Lodging

Airport phones: 307-326-8344, -8693 for Saratoga Aviation. Fuel: 100 and Jet. The town is within walking distance, and all motels and inns listed pickup from the airport with prior arrangements. Saratoga Aviation rents cars for $65-110 per day.

The recently remodeled Saratoga Inn is the largest public facility in Saratoga. It offers **golf**, **tennis**, **pool**, outdoor **hot springs**, and **fishing** from the trout-rich river that passes through the golf course. The new owners have added a **microbrewery** that creates award-winning microbrew.

The Hotel Wolf, located in the center of town, was built in 1893 and served time as a stagecoach stop. It has a restaurant and historic bar, and the rooms vary greatly in style and price. The Hood House is a 100 year-old restored Victorian B&B, full of antiques and quite unusual for a town of Saratoga's size.

© 2001 by ALTA Research

Saratoga **WY - SARA**

Brush Creek Ranch, 20 miles southeast, is a working cattle ranch with 1900-vintage lodge and log cabins for visitors. Lots of outdoor activities are offered, or you can simply join in and do the chores. The Medicine Bow Lodge is a country neighbor, also with historic lodge and cabins. It specializes in diverse week-long **western experiences**, appropriate for all seasons of the year. The Hacienda Motel may be the only place in town with air conditioning.

Features
B Full breakfast
C Continental
H Historic
J Jacuzzi
P Pets OK
R Restaurant
S Swim pool
T Transportation

Brush Creek Ranch	**BHJPT**	$460	800-RANCHWY
Hacienda Motel (airport)	**PT**	$60-76	307-326-5751
Hood House B&B	**BHPT**	$75	307-326-8901
Medicine Bow Lodge	**BHJT**	$104	307-326-5439
Rivera Lodge	**PT**	$42-173	307-326-5651
Saratoga Inn	**HJPRT**	$106-236	800-594-0178
Silver Moon Motel	**PT**	$33-46	307-326-5974
Wolf Hotel	**HT**	$51-97	307-326-5525

Saratoga and Area

The Saratoga area was visited by white men in the early 1800s when western pathfinders and mountain men camped by the North Platte River. Until 1860, only a handful of white settlers had been held by the beautiful vistas, tremendous game herds, exceptional fishing, and hot springs. But soon after, the land was noticed by settlers who saw its potential for ranching.

The **hot springs** were first developed by William H. Caldwell, who realized their potential while hunting meat for the railroaders. He built a bathhouse with two tubs and began to sell baths, hot meals, and space for travelers to sleep. Now called the **Hobo Hot Springs**, the facilities are free and open around the clock.

© 2001 by ALTA Research

Today, sleepy Saratoga continues Caldwell's tradition by catering to vacationing travelers. Home of an exclusive resort, **country club resort**, and a variety of lodging, Saratoga is the right place to kick back and get away from the rat race.

Viewed from above, the town's most prominent landmark is the Saratoga Inn's bright green **golf course**. Prices range from $13 for 9 holes to $19 for 18. The Inn's room rental includes free **swimming pool**, **hot springs**, **tennis courts**, and **private fishing**.

Saratoga is a small town, and in-town recreation is somewhat limited. You can visit the free **Saratoga Museum**, 307-326-5511, which is located in the original 1915 Union Pacific Railroad Depot. The exhibits include a country store, pioneer home, tie hacking equipment, and land office. Or, visit the 77 year-old Saratoga National **Fish Hatchery**. You can even browse an **art gallery**; Saratoga has several arty shops to visit.

Writing post cards at the Saratoga Inn, "Wish you were here . . ."

Saratoga **WY - SARA**

The old mining community of Encampment is 20 miles south. The town owes its heritage to an English sheepherder who discovered raw copper ore in 1897. The **Grand Encampment Museum** is open afternoons, 307-327-5308.

The Medicine Bow National Forest, 307-745-8971, circles the valley to the south. It contains 1,093,532 acres and includes parts of three mountain ranges: the Laramie, Medicine Bow, and Sierra Madre. Elevations range from 5000 to 12,013 feet. The area contains four wilderness areas: Savage River, Platte River, Huston Park, and Encampment River. These unpopulated areas can provide a leisurely day's drive or rugged **wilderness experience**.

The riverside **Hobo Hot Springs** is a public facility with free hot soaking pool, and a warm **swimming pool** with a fee. Camping is not allowed. Although the structures are in need of maintenance, the hot water sure feels good.

Hobo Hot Springs: river and hot pool are free, swim in the pool for a fee

If you landed at Saratoga solely for fuel, why not take an extra hour and treat yourself to a soak. It's less than ten minutes from airport to ecstasy. Grab a swim suit and walk east across the highway two blocks to Greenwood Street. Hang a left and walk four or five blocks north to the pool and springs.

After dinner, there's bound to be **music** and **dancing** at the Wolf Hotel or Saratoga Inn Fireside Lounge.

Perhaps Saratoga's greatest source of pride is its offering to the sportsman. From Saratoga, upstream 70 miles to the Colorado state line, the Upper Platte River is the premier **trout stream** in the state. This stretch contains 3200 catchable trout per mile. The most effective way to get your share is to float the river. The longest possible trip, from Six Mile Gap to Ft. Steele, takes almost three days.

Count on spending about $375 per day per boat for float trips; half days are $250-300. Great Rocky Mountain Outfitters, 307-326-8750, offers **float and wade trips**. See the Chamber web site for a complete list of outfitters.

Between summer and winter is **big game hunting** time. Local folks brag about opportunities for bagging the likes of deer, elk, antelope, bear, mountain lion, rabbits, game birds, big horn sheep, and even an occasional moose. Winter brings limitless space for **cross-country skiing** and **snowmobiling**.

Events and Festivals

Jan, mid	Ice Fishing Derby
June, mid	Woodchoppers Jamboree & Rodeo
July, early	Art Festival
Aug, early	Carnival Crazy Daze
Aug, late	Sierra Madre Muzzle Loader Rendezvous
Aug, late	Microbrew Competition

© 2001 by ALTA Research

WY - THER

Thermopolis

Hot Springs County Airport

Town	Thermopolis **ID** THP
Coord	N43-39.50, W108-12.79
Elev	4580 ft, S end high, 2.5%
Runway	1 - 19, 4800 x 100' asphalt
Freq	CTAF-122.8, VASI-122.8-3x
Charts	Cheyenne sect, L9, Lo7

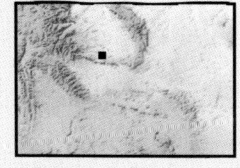

CAUTION: *Airport information not for navigational use. Hills both sides. Night ops not recommended. Ultralights, RC airplanes, deer. Land on 19 unless tailwinds are above 15 kts.*

> ***Thermopolis**, home of the world's largest hot springs complex – three large hotwater play centers. Fishing and hunting. Rich in Native American, outlaw, geological history. $60-500*
> *CofC: 800-786-6772, www.thermopolis.com*

Airport Description

The airport is just a mile north of town. Although airport elevation is higher than the valley floor, nearby land surface rises even higher above the airport on either side of the runway. This can be a hazard at night for pilots unfamiliar with the area.

If tail winds are less than 15 knots, traffic lands south and departs north. This is compatible with runway gradient, and minimizes noise over the town. While on downwind for Runway 19, sneak a peek at the expansive hot springs works below.

Airport Services, Transportation, and Lodging

Airport phone: 307-864-2488. Fuel: 100LL. $10 landing fee, unless fuel purchased. G&W Aviation has a mechanic on duty, providing minor and major repairs. They keep the runway clear in winter, and have a hangar for overnight parking. Rental cars cost $30-up per day from their Rent-a-Wreck franchise. In town, most everything is within walking distance.

Lodging is plentiful and affordable in Thermopolis. The Holiday Inn of The Waters is located at Hot Springs Park. The Holiday is relatively upscale, offering amenities as: masseuse on duty, exercise room, and outside mineral Jacuzzi and pool. Check out their Trophy Bar.

Features
B Full breakfast
C Continental
H Historic
J Jacuzzi
P Pets OK
R Restaurant
S Swim pool
T Transportation

© 2001 by ALTA Research

Thermopolis   **WY - THER**

Cactus Inn	PT	$25-60	877-612-7811
El Rancho Motel	T	$40-60	800-283-2777
Elk Antler Inn	PT	$22-59	800-320-4028
High Island Ranch	BT	$400-500	307-864-5389
Holiday Inn	PRS	$70-150	307-864-3131
Hot Springs Motel	P	$38-50	307-864-2303
Quality Inn & Suites	CH	64-134	888-919-9009
Moonlighter Best Western	CJRST	$42-74	307-864-2321
Roundtop Mtn. Motel	PT	$38-74	307-864-3126
Super 8 Motel	CJPST	$48-90	307-864-5515

Although overnight camping is no longer allowed at the airport, pilots can pitch a tent one mile north at the Fountain of Youth RV Park, 307-864-3265. Campers enjoy a 235-foot hot springs pool, refreshed with 1.4 million gallons daily. The overnight cost is around $20 for two. The owner, a pilot, provides free pickup from the airport.

Quality Inn & Suites, a touch of history near the springs

The High Island Ranch and Cattle Company is a 41,000-acre working cattle ranch that lets you take part in a 35-mile cattle drive – you sign on as a dude and leave as a cowpuncher. Sign up for their "Period Drive," and they loan you cowboy or cowgirl clothing from the 1800s. What fun!

Ride hard or kick your heels up and dream, your option at High Island. They offer trail rides, outfitting, and fly fishing excursions. Cost is $1300-up each per week, plus optional gratuity. The fee includes transportation from the airport.

Thermopolis

Originally, Thermopolis' **hot springs** were included in the Shoshoni Indian Reservation of 1868. Later, the Arapahoe Indians were also moved to the reservation. Members of both tribes visited the springs, which they called Bah-gue-wana, or

Round-up at High Island Ranch　　　　Photo: Karen Cummings

Thermopolis

Smoking Water. They looked upon the Big Spring with awe and respect, using it to heal their ills.

White Man's history officially began in 1897 when the U.S. government purchased a ten mile square tract of land from the Indians. It contained Big Spring, the world's largest hot spring. A square mile containing the spring was presented to the State of Wyoming. The treaty, signed by Chief Washakie, specified that a portion of the water from the spring be held forever free to the people.

Today, visitors can select among several options. The Tepee Pools offer indoor and outdoor facilities, including a 161-foot indoor slide and 272-foot outdoor slide. The Star Plunge is packed with attractions, including three **water slides** (up to 500 feet long), **swimming pools**, **hot pools**, **Jacuzzis**, **high dive**, **volley ball court**, and more.

Star Plunge facility

In honoring Chief Washakie's agreement, the State of Wyoming provides free access at a third facility for 15-20 minute intervals. I wonder if the rigid access procedure is what the Chief had in mind. To the state's credit, however, the facility is spotless and you can rent bathing suits and towels for a nominal charge.

In between soaks, you can play a round of **golf** by the airport, 307-864-5294, or visit the **museums**. The Historical Museum and Cultural Center, 864-5183, is at 700 Broadway. This museum contains the cherrywood bar from the Hole-in-the-Wall Saloon, and exhibits that contain Native American, pioneer, and outlaw artifacts. Kids and adults may enjoy the new Old West Wax Museum, 864-9396.

The **Wyoming Dinosaur Center**, a recent addition to Thermopolis, has dinosaur-related fossils and dynamic displays.

Free, state-operated soaking area, with cascades of calcium deposits

Thermopolis

Entry is $6, and less for children. You can dig at the fossil site, or tour the dig for a fee. Call 800-455-3466 for information.

Need a quick diversion for the kids? Hop in the car and try locating the **buffalo herd** on the range land to the east. Take any of three paved roads that leave the east edge of Hot Springs Park.

To the west of Thermopolis at Legend Rock in Hot Springs State Park, 307-864-2176, you can view one of the most impressive **petroglyph displays** in Wyoming. A recent survey estimates 283 pictures on 93 rock panels, the oldest of which dates back 2000 years. To get to the site, you take Highway 120 west.

Those who like to take life slow will appreciate the laid-back pace around the spas. After a soak or two in the pools, may I suggest a **massage** from the local talent in town. If there's still an hour of sun, you have time for a leisurely drive through the Wind River Canyon.

The 15-mile stretch through the Wind River Canyon to Boysen Lake is a pleasant **drive**. The drive south on Highway 20, takes you past millions of years of textured geological history. The play of light on multicolored rocks and foliage is ever changing. After passing old railroad tunnels carved out of solid rock, you reach Boysen Lake. The lake has **camping**, **fishing**, and **boating**.

North of town in Kirby on Highway 20, you can order a burger the size of a plate at Butch's. On the way, stop and see Domhoff Pottery. Back in Thermopolis, the Legion Supper Club at the golf course on Airport Hill is known for **fine dining** – perhaps the best in town. Pumpernicks or Grannies are also safe bets for a **good meal**. Little Wrangler and Sideboard are favorite **breakfast** places in town.

With a warm glow in your tummy and not a tense muscle on your frame, it's OK to head back to your room for a snooze at any

© 2001 by **ALTA** Research

time. Just turn back the sheets, hit the sack, and allow yourself to drift off as the breeze flutters the curtains by the window.

Good **fishing** holes can be found in and around Thermopolis in the Big Horn River. The official analysis shows 1350 trout per mile. Some fishermen use rubber rafts and jon boats to increase their chances of landing a lunker. Starting points are at the Wedding of the Waters or at Harvey Public Fishing Area.

The High Island Ranch specializes in **fly fishing** trips and **big game hunts** for deer, elk, antelope, moose, and sheep. The owners claim hunting success rates of 80% to 100%. Contact the Wyoming Outfitters Association and Guides at 307-527-7453 for names of other outfitters.

Events

The Gift of the Waters **Pageant** in early August celebrates the sale of the Hot Springs by the Shoshone and Arapahoe Indians. The play was written in 1925 and has been performed since 1950. The event includes **Indian dancers** in authentic dress.

Several days later, Outlaw Trail, Inc. puts on their Wyoming **Trail Ride**. One hundred riders ride 100 miles, following the trails of Butch Cassidy and the Sundance Kid from Hole-in-the-Wall country to Thermopolis. The event lasts six days and costs about $700 for horse and meals.

A list of other significant events follows. Call the Chamber of Commerce, or see their web site for latest details.

June, mid	Thermopolis Indian Days
June, end	Wild West Weekend / Bits, Spurs, Saddles Show
Aug, early	Gift of the Waters Pageant
Aug, mid	Outlaw Trail Ride
Aug, mid	Ranch Days Rodeo
Aug, end	Labor Day Golf Tournament

© 2001 by **ALTA** Research

Index

aerial tram,	OR-WALL	Bigfork,	MT-BIGF
airpark,	ID-LAVA	**boating,**	OR-WALL
	OR-SIST	OR-HOOD	OR-NEWP
air show,	ID-SILV	OR-TILL	WA-FRID
	OR-TILL	WA-LOPE	WA-ORCA
		WA-ROCH	WY-THER
Alpine,	WY-ALPI	Boyer YL Ranch,	WY-BOYE
Alvord Desert,	OR-ALVO	Bruce Meadows,	ID-BRUC
amusement park,	ID-SILV	**camp at airport,**	ID-BIGC
	OR-SEAS		ID-BHUC
		ID-DIXT	ID-DIXU
aquarium,	OR-NEWP	ID-CHAM	ID-JOHN
	OR-SEAS	ID-LAVA	ID-MOOS
		ID-MURP	ID-SMIL
art,	MT-BIGF	ID-THOM	ID-WARM
OR-BAND	OR-HOOD	MT-BENC	MT-BIGF
OR-TROU	OR-WALL	MT-SCHA	MT-SPOT
WA-LEAV	WA-ORCA	OR-ALVO	OR-DIAM
		OR-MEMA	OR-MINA
Ashland,	OR-ASHL	OR-OWYH	OR-PINE
	OR-PINE	OR-SIST	OR-WALL
		WA-LEAV	WA-ORCA
balloon rides,	OR-WALL	WA-PACK	WA-PASA
		WA-RANG	WA-TIET
Bandon,	OR-BAND	WY-ALPI	WY-PINE
Benchmark,	MT-BENC	**canyons,**	OR-MEMA
Bend,	OR-SUNR	Chamberlain,	ID-CHAM
Big Creek,	ID-BIGC	Chico Hot Springs,	MT-CHIC

© 2001 by **ALTA** Research

Crescent Lake Junction, OR-DIAM

Crescent Lake State Park, OR-DIAM

cross-country skiing, *see* **skiing**

cruises, OR-HOOD
OR-NEWP OR-TILL
WA-FRID WA-ORCA

dogsled rides, MT-CHIC

Diamond Peak Wilderness, OR-DIAM

Dixie Town, ID-DIXT

Dixie USFS, ID-DIXU

dude ranch, OR-MINA
OR-PINE
ID-FLYB ID-JOHN
ID-SULF MT-BENC
MT-SPOT WY-BOYE
WY-SARA WY-THER

Eagle Cap Wilderness, OR-HORS
OR-MINA OR-WALL

Enterprise, OR-WALL

Evergreen Aviation Museum, OR-MCMI

fishing, ocean, OR-BAND
OR-NEWP
OR-TILL OR-SEAS
WA-FRID WA-LOPE
WA-ORCA WA-ROCH

fishing, stream / lake,

ID-BIGC ID-BRUC
ID-CHAM ID-DIXT
ID-DIXU ID-FLYB
ID-JOHN ID-LAVA
ID-MOOS ID-MURP
ID-SMIL ID-STAN
ID-SULF ID-THOM
ID-WARM MT-BIGF
MT-CHIC MT-SCHA
MT-SPOT OR-DIAM
OR-HOOD OR-HORS
OR-MINA OR-OWYH
OR-SUNR OR-WALL
WA-LEAV WA-PASA
WA-RANG WA-TIET
WY-ALPI WY-BOYE
WY-PINE WY-SARA
WY-THER

Flying B Ranch, ID-FLYB

Flying M Ranch, OR-FLYM

food near airport, ID-BIGC
ID-DIXT
ID-FLYB ID-SILV
ID-SMIL ID-SULF
MT-BENC MF-SPOT

© 2001 by **ALTA** Research

Index-4

OR-FLYM	OR-MINA
OR-SIST	OR-SUNR
OR-TILL	WA-FRID
WA-ROCH	WA-FRID

Friday Harbor, WA-FRID

Galena, ID-SMIL

ghost towns,
| ID-STAN | ID-SMIL |
| OR-DIXT | OR-DIXU |

glider rides, ID-SILV
OR-MCMI

golf, ID-LAVA MT-BIGF
OR-BAND	OR-HOOD
OR-SEAS	OR-SIST
OR-SUNR	OR-ASHL
WA-LEAV	WY-PINE
WY-SARA	WY-THER

hay rides, ID-LAVA
OR-FLYM OR-PINE

hiking,
see **wilderness hiking**

Hood River, OR-HOOD

Horse Ranch, OR-HORS

horseback riding, ID-BIGC
ID-BRUC	ID-DIXT
ID-FLYB	ID-LAVA
ID-MURP	ID-SMIL
ID-STAN	ID-SULF
MT-BENC	MT-CHIC
MT-SPOT	OR-ALVO
OR-BAND	OR-FLYM
OR-HOOD	OR-MINA
OR-PINE	OR-SEAS
OR-SIST	OR-SULF
OR-SUNR	OR-WALL
WY-BOYE	WY-PINE
WY-SARA	WY-THER

horse-drawn rides, ID-LAVA
OR-FLYM
OR-PINE

hot air balloon rides,
OR-WALL

Hot Springs, MT-HOTS

hot springs, ID-JOHN
ID-LAVA
ID-MURP	ID-STAN
ID-THOM	ID-WARM
MT-CHIC	MT-HOTS
WY-SARA	WY-THER

hunting, ID-BIGC
ID-CHAM	ID-FLYB
ID-LAVA	ID-MURP
ID-STAN	ID-SULF
ID-THOM	OR-MINA
OR-WALL	MT-BENC
MT-SPOT	WY-PINE
WY-SARA	WY-THER

© 2001 by **ALTA** Research

Indian museum, OR-WALL	Murphy Hot Springs, ID-MURP
Indian show, ID-LAVA / WY-THER	**museum, air,** OR-MCMI / OR-TILL
island, WA-FRID / WA-LOPE / WA-ORCH / WA-ROCH	**museum, history,** ID-LAVA / OR-HOOD / OR-MCMI / OR-NEWP / OR-TILL / OR-WALL / WY-SARA
Johnson Creek, ID-JOHN	**museum, Indian,** OR-WALL
Joseph, OR-WALL	Newport, OR-NEWP
kitesurfing, OR-HOOD	**ocean,** OR-BAND / OR-NEWP / OR-SEAS / OR-TILL / WA-FRID / WA-LOPE / WA-ORCA / WA-ROCH
Lava Hot Springs, ID-LAVA	
Leavenworth, WA-LEAV	
Lopez Island, WA-LOPE	
Lost River Resort Airport, WA-PASA	Orcas Island, WA-ORCA
McMinnville, OR-MCMI	Owyhee Reservoir, OR-OWYH
Memaloose, OR-MEMA	**pack trips,** ID-BIGC / ID-CHAM / ID-FLYB / ID-JOHN / ID-STAN / ID-SULF / MT-BENC / MT-SPOT / OR-MINA / OR-WALL / WY-BOYE / WY-PINE / WY-SARA / WY-THER
Minam, OR-MINA	
mine, historic ID-BIGC / ID-JOHN / ID-SMIL / ID-STAN / WY-SARA / WY-THER	
Moose Creek, ID-MOOS	
Mount Rainier, WA-RANG	

© 2001 by **ALTA** Research

Index-6

Flyer's Recreation Guide

Packwood,	WA-PACK
paraglide,	OR-HOOD
Pasayten Wilderness,	WA-PASA
Pinedale,	WY-PINE
Pinehurst,	OR-PINE
Pocatello,	ID-LAVA

rafting, ID-STAN
 OR-ASHL
 OR-HOOD OR-PINE
 OR-SUNR OR-WALL
 WA-LEAV WY-ALPI

ranch, *see* **dude ranch**

Ranger Creek, WA-RANG

resort, ID-BIGC
 ID-FLYB
 ID-JOHN ID-STAN
 ID-SULF ID-THOM
 MT-BENC MT-BIGF
 MT-CHIC MT-SPOT
 OR-MEMA OR-SUNR
 WA-ORCA WA-ROCH
 WY-BOYE WY-SARA

Roche Harbor, WA-ROCH

sailing, MT-BIGF
 OR-HOOD WA-FRID
 WA-ORCA WA-ROCH

San Juan Islands, WA-FRID
 WA-LOPE
 WA-ORCA WA-ROCH

Santiam Junction, OR-SANT

Saratoga, WY-SARA

Savery Creek Ranch, WY-BOYE

Schafer, MT-SCHA

scuba diving, WA-FRID
 WA-ROCH
 WA-LOPA

Seaside, OR-SEAS

Silverwood, ID-SILV

Sisters, OR-SIST

skeet shoot, ID-FLYB

skiing, cross-country,
 ID-BIGC
 ID-JOHN ID-LAVA
 ID-STAN MT-CHIC
 OR-ASHL OR-SIST
 OR-SUNR WY-ALPI
 WY-SARA WY-PINE

© 2001 by **ALTA** Research

skiing, down-hill, ID-LAVA
OR-ASHL
OR-HOOD OR-SUNR

Smiley Creek, ID-SMIL

snowmobile rental,
WY-PINE WY-SARA

soaring, ID-SILV
OR-MCMI

Spotted Bear, MT-SPOT

Stanley, ID-STAN

Sulfur Springs Ranch,
OR-SULF

Sunriver, OR-SUNR

theater, MT-BIGF
OR-ASHL
OR-BAND OR-HOOD
OR-SEAS WA-ORCA

Thermopolis, WY-THER

Thomas Creek, ID-THOM

Tieton, WA-TIET

Tillamook, OR-TILL

train, historic, ID-SILV
OR-HOOD

tramway, OR-WALL

Troutdale, OR-TROU

vineyard, OR-HOOD
OR-MCMI

Wallowa Recreation Area,
OR-WALL

Warm Springs, ID-WARM

water slide, ID-LAVA
WY-THER

whitewater trips,
see **rafting**

wilderness hiking,

ID-BIGC	ID-BRUC
ID-CHAM	ID-DIXT
ID-DIXU	ID-FLYB
ID-JOHN	ID-MOOS
ID-SMIL	ID-STAN
ID-SULF	ID-THOM
ID-WARM	MT-BENC
MT-SCHA	MT-SPOT
OR-DIAM	OR-HORS
OR-MCMA	OR-MINA
OR-OWYH	OR-SANT
OR-SIST	OR-WALL
WA-LEAV	WA-PASA
WY-ALPI	WY-BOYE
WY-PINE	WY-SARA

windsurfing, OR-HOOD

© 2001 by **ALTA** Research

© 2001 by ALTA Research